97-CTS-4

# ASSESSING METHODS OF REMOVING METALS FROM WASTEWATER: THE EFFECT OF FERRIC CHLORIDE ADDITION ON METAL SPECIATION AND REMOVAL

by:
David L. Sedlak
A. Christianne Ridge
William W. Bedsworth
University of California, Berkeley

*2004*

WERF

The Water Environment Research Foundation, a not-for-profit organization, funds and manages water quality research for its subscribers through a diverse public-private partnership between municipal utilities, corporations, academia, industry, and the federal government. WERF subscribers include municipal and regional water and wastewater utilities, industrial corporations, environmental engineering firms, and others that share a commitment to cost-effective water quality solutions. WERF is dedicated to advancing science and technology addressing water quality issues as they impact water resources, the atmosphere, the lands, and quality of life.

For more information, contact:
Water Environment Research Foundation
635 Slaters Lane, Suite 300
Alexandria, VA 22314-1177
Tel: (703) 684-2470
Fax: (703) 299-0742
www.werf.org
werf@werf.org

This report was co-published by the following organizations. For non-subscriber sales information, contact:

IWA Publishing
Alliance House, 12 Caxton Street
London SW1H 0QS, United Kingdom
Tel: +44 (0) 20 7654 5500
Fax: +44 (0) 20 7654 5555
www.iwapublishing.com
publications@iwap.co.uk


Library of Congress Catalog Card Number: 2003107281

IWAP ISBN: 1-84339-694-7



University of California, Berkeley

# ACKNOWLEDGMENTS

The project team would like to acknowledge and thank Patricia McGovern and Steve McDonald of Carollo Engineers for their suggestions on the design and implementation of the bench-scale experiments. The team also thanks Joshua Stolaroff, Laurel Schaider, and Helen Hsu for their assistance in the bench-scale experiments and the staff of the Sacramento Regional and Millbrae wastewater treatment plants for their cooperation and the use of their laboratories.

## Report Preparation

**Principal Investigator:**
David L. Sedlak, Ph.D.
*University of California, Berkeley*

**Project Team:**
A. Christianne Ridge, M.S.
*University of California, Berkeley*

William W. Bedsworth, Ph.D.
*University of California, Berkeley*

## Project Subcommittee

Matthew Gerhardt, Ph.D., P.E., DEE, chair
*Brown and Caldwell, Inc.*

Michael Aitken, Ph.D.
*University of North Carolina, Chapel Hill*

C.P. Huang, Ph.D., P.E.
*University of Delaware*

Yi-Tin Wang, Ph.D.. P.E.
*University of Kentucky*

## Water Environment Research Foundation Staff

**Director of Research:** Daniel M. Woltering, Ph.D.
**Director of Technical Services:** Amit Pramanik, Ph.D.

# ABSTRACT AND BENEFITS

**Abstract:**

The presence of cationic pollutant metals in municipal wastewater effluent is a concern because stringent discharge requirements cannot always be met with conventional treatment methods. Attempts to improve metal removal are often unsuccessful because a significant fraction of the cationic metals are complexed by the synthetic chelating agent ethylenediamine tetraacetic acid (EDTA).

To identify practical approaches for improving metal removal, an analytical method for measuring metal-EDTA complexes was used to survey metal speciation at a series of wastewater treatment plants. Following these analyses, bench-scale experiments were conducted. The survey data indicated that pollutant metal-EDTA complexes account for a significant fraction of the dissolved metals in wastewater. The bench-scale studies indicated that ferric chloride ($FeCl_3$) addition improves the removal of Cu and Zn by approximately 20%.

To test the results of the bench-scale experiments, a full-scale experiment was conducted by interrupting chemical addition at a municipal wastewater treatment plant that normally adds $FeCl_3$ during primary treatment. Results indicated that $FeCl_3$ addition had a slight impact on metal speciation but no effect on metals removal. The lack of an effect was attributed to changes in metal speciation that occurred during primary treatment irrespective of $FeCl_3$ addition.

**Benefits:**

- Evaluates approaches for improving the removal of cationic metals by chemical addition during primary wastewater treatment.
- Assesses the role of synthetic chelating agents and pH in the speciation and removal of cationic metals during wastewater treatment.
- Provides a new analytical method for direct measurement of metal complexes with the synthetic chelating agent EDTA in municipal wastewater.
- Identifies changes in metal speciation that occur during primary wastewater treatment.

**Keywords:** Copper, ethylenediamine tetraacetic acid (EDTA), $FeCl_3$, metal speciation, zinc

# TABLE OF CONTENTS

# LIST OF TABLES

# LIST OF FIGURES

# LIST OF ACRONYMS

| | |
|---|---|
| CRCP | Chelating-resin column partitioning |
| DOC | Dissolved organic carbon |
| EDTA | Ethylenediamine tetraacetic acid |
| GFAAS | Graphite furnace atomic absorbance |
| GR | Grand Rapids |
| HPLC | High performance liquid chromatography |
| NonFeEDTA | The sum of all EDTA that is not complexed by Fe |
| RAS | Return activated sludge |
| SESF | Southeast San Francisco |
| SJSC | San Jose/Santa Clara |
| SR | Sacramento Regional |
| TSS | Total suspended solids |
| WWTP | Wastewater treatment plant |

# EXECUTIVE SUMMARY

The removal of cationic pollutant metals during municipal wastewater treatment is controlled by the interaction of dissolved metals with surfaces and microorganisms. Available data, summarized in a previous WERF report (*Assessing Methods of Removing Metals from Wastewater: A Review of Data and Methodologies*; 2000), indicate that these interactions are affected by metal speciation. In particular, the presence of the synthetic chelating agent ethylenediamine tetraacetic acid (EDTA) results in the formation of negatively charged metal species that have a low tendency to adsorb onto surfaces. Due to the presence of metal-EDTA complexes in secondary effluent, some wastewater treatment plants are unable to achieve stringent discharge limits being enforced in locations where wastewater effluent is discharged into sensitive habitats. To identify cost-effective approaches for improving the removal of cationic metals, a new analytical method for measuring metal speciation was used to survey metal speciation at four wastewater treatment plants; then a series of bench-scale experiments and full-scale manipulations were conducted to assess the potential for improving metal removal by chemical addition during primary treatment.

The new analytical method developed as part of this project is capable of discriminating between the different pollutant metal-EDTA complexes at the concentrations encountered in municipal wastewater. Application of the method indicated that Cu and Zn complexes with EDTA are present in primary effluent and that $ZnEDTA^{2-}$ accounts for a significant fraction of the dissolved Zn discharged by wastewater treatment plants.

To assess the possibility of improving metal removal by manipulation of speciation, bench-scale experiments were conducted by subjecting raw sewage to a series of manipulations designed to simulate the processes that occur in a conventional wastewater treatment plant. To simulate primary treatment, sewage was settled and decanted into a second container where secondary treatment was simulated by mixing the supernatant with fresh return activated sludge and aerated. After aeration, the mixture was settled to simulate secondary clarification. Samples collected after simulated primary and secondary treatment were analyzed for dissolved and total metals, metal speciation and concentrations of different metal-EDTA complexes. To assess the effects of chemical addition during primary treatment, results from treatments in which $FeCl_3$ and/or HCl were added prior to simulated primary treatment were compared with controls. Addition of $FeCl_3$ resulted in a decrease of approximately 20% in the concentrations of Cu and Zn in the effluent. Addition of HCl, included in the study to assess the effect of acidity contributed by the $FeCl_3$, did not result in any consistent improvement in metal removal. Measurement of metal speciation indicated that the improvement in metal removal was partially attributable to the conversion of $CuEDTA^{2-}$ and $ZnEDTA^{2-}$ into FeEDTA complexes. Some of the additional metal removal was attributable to better settling of particles and stronger adsorption of metals onto particle surfaces.

To verify the findings of the bench-scale studies, a full-scale manipulation was conducted at the Millbrae (Calif.) municipal wastewater treatment plant. To assess the role of chemical addition in metal removal, samples were collected at two-day intervals during a three-month period in which $FeCl_3$ addition was interrupted and resumed. Results of the full-scale study indicated that $FeCl_3$ addition at a dose of 10 mg/L during primary treatment did not improve the removal of Cu and Zn. Small decreases in the concentrations of $CuEDTA^{2-}$ and $ZnEDTA^{2-}$ observed during the period in which $FeCl_3$ was added were overwhelmed by changes in metal speciation that occurred during primary treatment, irrespective of chemical addition. The changes in metal speciation during primary treatment were related to the presence of particles. If this process could be understood and controlled, it might be used to enhance the dissociation of pollutant metal-EDTA and thereby improve metal removal.

# CHAPTER 1.0

# INTRODUCTION

## 1.1. Background

Pollutant metals in municipal wastewater effluent are a concern because relatively low concentrations of metals can have adverse effects on humans and aquatic organisms. As a result of increasing concerns about these effects, discharge limits for pollutant metals have been decreasing in recent decades. Furthermore, efforts to control these discharges at the watershed level provide incentives to limit the mass of pollutant metals discharged by municipal wastewater treatment plants (WWTPs). Therefore, improvement in pollutant metal removal may be desirable, even if the concentrations of pollutant metals in wastewater effluent are already very low.

The concentrations of pollutant metals discharged by most WWTPs are difficult to reduce because they are optimized to meet other water quality objectives. Any modification in plant operating conditions must not compromise plant performance or decrease plant reliability. Approaches for improving metal removal also must be relatively inexpensive. As a result, feasible approaches for improving pollutant metal removal efficacy are limited to relatively small modifications to existing treatment processes or installation of cost-effective treatment systems with proven reliability. The purpose of this research project was to identify cost-effective approaches for achieving extremely low concentrations of metals in municipal wastewater effluents.

During the first phase of the project, issues related to pollutant metals in municipal wastewater effluents were examined. A WERF report on this phase of the project summarized the results of the review (WERF, 2000). In the review, five metals were identified that are frequently detected at concentrations near or above the most stringent discharge limits. Two of these, mercury and silver, are frequently detected in association with particles or colloids in wastewater effluents, while the remaining three, Cu, Ni, and Zn, are almost always in the dissolved phase in wastewater effluent. Improving metal removal efficiency for the particle-associated metals can be achieved through better control of particles (effluent filtration), but analysis of equilibrium speciation models indicates that because Cu, Ni, and Zn occur mainly in dissolved forms and exhibit a high affinity for organic ligands, they are difficult to remove.

Evaluation of the organic ligands likely to be present in municipal wastewater suggests that the synthetic chelating agent ethylenediamine tetraacetic acid (EDTA) is important to complexation of Cu, Ni, and Zn. The presence of complexes of these three pollutant metals with EDTA is a significant concern because previous research indicates that metal-EDTA complexes are nearly impossible to remove from municipal wastewater (Cheng et al. 1975). One possible approach for treating pollutant metal-EDTA complexes involves their conversion into FeEDTA complexes. After the pollutant metals are released from the EDTA complexes, they may be removed during conventional treatment. Results of equilibrium speciation models suggested that adjusting pH to values below 7.0 is a viable approach for converting Cu, Ni and Zn complexes with EDTA into FeEDTA complexes. In addition, measurements of concentrations of FeEDTA complexes made at full-scale WWTPs (Kari and Giger, 1996) suggest that adding $FeCl_3$ during primary treatment also results in conversion of pollutant metal-EDTA complexes into FeEDTA complexes.

To assess the potential for improving the removal of Cu, Ni, and Zn by manipulation of metal speciation, better information was needed on the speciation of metals in wastewater and the role of chemical addition on metal speciation and removal. To achieve these goals, a new analytical technique was developed for measuring the speciation of metal-EDTA complexes. The method was used to measure the speciation of metals at four wastewater treatment plants that employed different treatment technologies and that used different unit processes. After completing the analyses of metal speciation, bench-scale and full-scale studies were conducted to assess the effects on metal removal and speciation of adding $FeCl_3$ and/or HCl during primary treatment.

# CHAPTER 2.0

# SPECIATION OF METALS IN MUNICIPAL WASTEWATER

## 2.1 Background

As mentioned in Chapter 1.0, results of the literature review suggested that complexation of pollutant metals by EDTA plays an important role in metal removal by wastewater treatment plants. However, the ability to measure EDTA complexes was limited by the existing high performance liquid chromatography (HPLC) and gas chromatography (GC) techniques, which only measured total EDTA (Kari and Giger 1996; Nowack et al., 1996). Because visible light destroys EDTA complexed by Fe(III), the HPLC and GC methods can be modified to quantify FeEDTA complexes by measuring the difference between total EDTA before and after photolysis of samples. $NiEDTA^{2-}$ also can be measured with these methods by heating the sample to take advantage of the slow ligand exchange kinetics of Ni(II) (Nowack et al., 1996). The difference between the total EDTA concentration and the concentration remaining after photolysis, which is referred to here as nonFeEDTA complexes, provides evidence for the existence of pollutant metal-EDTA complexes. However, it does not provide information on which complexes are present.

To confirm the hypotheses about the presence of $CuEDTA^{2-}$ and $ZnEDTA^{2-}$ in municipal wastewater, an analytical method was needed that could directly quantify the pollutant metal-EDTA complexes. During this phase of the project, a new method was developed for quantifying metal-EDTA complexes in wastewater. A brief description of the method is included in this chapter, followed by results of measurements made in samples collected from four different WWTPs. Full details of the analytical method have been published in a peer-reviewed scientific paper (Bedsworth and Sedlak, 2001).

## 2.2 Direct Detection of Metal-EDTA Complexes by HPLC

Several different analytical methods have been described for the direct measurement of metal-EDTA complexes (Buchberger et al., 1991; Ye and Lucy, 1995). However, the available methods suffer from an unacceptable amount of interference in the presence of organic matter in wastewater effluent. In the new method developed for this study, the HPLC conditions described by Ye and Lucy (1995) were used to separate the metal-EDTA complexes. As part of this approach, the HPLC column was loaded with cetyltrimethylammonium bromide, an ion-pairing agent. This treatment provided ion-exchange sites for the negatively charged metal-EDTA complexes, which were then eluted with an aqueous mobile phase containing sodium sulfate buffered to pH 7 with 4-(2-hydroxyethyl)-1-piperazineethanesulfonic acid (HEPES). After separation, a post-column reaction was used to convert all of the EDTA complexes into FeEDTA complexes, which were detected by UV absorbance at 258 nm.

The detection limit of the method, based on a signal to noise ratio of 3, is 60-80 nM (5-8 ng for a 250-µL injection). This detection limit is in the same range as previous UV absorbance methods for Cu and Pb complexes with EDTA (Buchberger et al., 1991) but represents nearly an order of magnitude improvement in sensitivity for Co, Cd, and Zn complexes with EDTA. Although Co, Cd, or Pb complexes with EDTA in municipal wastewater were not expected, these metal-EDTA complexes were included in the method because they may be present in industrial wastewater or contaminated groundwater (Means et al., 1978; Cleveland and Rees, 1981).

After developing the method to quantify $CuEDTA^{2-}$ and $ZnEDTA^{2-}$, additional studies were performed to assess potential artifacts that could occur during analysis of wastewater effluents. For example, trace contamination of the HPLC system could result in the formation of $CuEDTA^{2-}$ or $ZnEDTA^{2-}$ from weaker EDTA complexes (complexes with lower conditional stability constants) during analysis. To minimize the potential for formation of these complexes, the entire HPLC system was passivated by passing dilute nitric acid through all the tubing and valves. In addition, a metal-free end-capped HPLC stationary phase (Supelco Discovery) was used to minimize active sites and metal contamination. To test the potential for formation of $CuEDTA^{2-}$ and $ZnEDTA^{2-}$ during analysis, solutions of $Na_2H_2EDTA$ were injected at concentrations as high as 1 mM. $Na_2H_2EDTA$ was used for these experiments because it represents a worst case scenario, since all other metal-EDTA salts ($CaEDTA^{2-}$) should be less reactive with trace metals. In all experiments, formation metal-EDTA complexes (complexes with Cu, Co, Cd, Pb, or Zn) accounted for less than 5% of the $Na_2EDTA^{2-}$ injected.

A more realistic test of the potential for artifacts was obtained by injecting $CaEDTA^{2-}$ because any EDTA present in wastewater should be complexed by a metal other than sodium. When 1 □M $CaEDTA^{2-}$ was injected in deionized water, a broad $CaEDTA^{2-}$ peak eluted, and no peaks attributable to $ZnEDTA^{2-}$ or $CuEDTA^{2-}$ were observed. When approximately 1 □M of $CaEDTA^{2-}$ was added to municipal wastewater effluent containing only FeEDTA complexes, a peak attributable to $CaEDTA^{2-}$ appeared prior to $ZnEDTA^{2-}$. The $ZnEDTA^{2-}$ peak was expected if some of the added $CaEDTA^{2-}$ was converted into $ZnEDTA^{2-}$ when the $CaEDTA^{2-}$ was added to the sample (equilibrium predictions suggest that $CaEDTA^{2-}$ will not be stable in the presence of free Zn). The $CaEDTA^{2-}$ peak continued to grow when additional $CaEDTA^{2-}$ was added; however, recovery of $CaEDTA^{2-}$ was only around 50%. The $ZnEDTA^{2-}$ peak did not increase after the EDTA concentration exceeded the Zn concentration in the sample. That incomplete recovery of $CaEDTA^{2-}$ may have been attributable to formation of a complex such as $FeEDTA^{-}$, which should also be more stable than $CaEDTA^{2-}$ under these conditions. These results indicate that even if $CaEDTA^{2-}$ were present, it would not be converted into other species to an appreciable degree during analysis.

To further assess the new analytical method, spike recovery experiments were performed by addition of $CuEDTA^{2-}$ and $ZnEDTA^{2-}$ to wastewater samples. Results of these experiments indicated that recovery of the EDTA complexes was nearly quantitative (mean recovery was 93% for $ZnEDTA^{2-}$ [n=22] and 90% for $CuEDTA^{2-}$ [n=12]). Although chromatograms from wastewater effluent samples exhibited more noise than deionized water standards, interfering peaks usually eluted prior to the metal-EDTA complexes. Using the two HPLC techniques (reversed-phase ion-pair chromatography and ion exchange chromatography with post-column conversion), it was possible to separate and quantify all of the major EDTA species expected in wastewater effluents. Application of the HPLC method to raw sewage and primary effluent was more problematic because interfering peaks attributable to organic matter often were encountered.

## 2.3 Measurement of Metal Speciation in Municipal Wastewater

### 2.3.1 Sample Locations and Methods

To assess the importance of metal-EDTA complexes in the speciation of metals in municipal wastewater, metal speciation in samples from four municipal wastewater treatment plants (WWTPs) were collected and analyzed (Table 2-1). The plants were chosen to represent a range of plant design configurations and effluent characteristics typical of WWTPs in the United States. Two oxygen-activated sludge treatment plants (Southeast San Francisco and Baltimore) were included to investigate the potential effects of temporary pH drops during secondary treatment. Samples were collected once from the Grand Rapids and Baltimore WWTPs and twice from the San Jose Santa Clara (SJSC) WWTP. Complete analysis of speciation was performed on one influent and two effluent samples from Southeast San Francisco (SESF). In addition, metal-EDTA complexes were analyzed in effluent samples collected on two other dates.

**Table 2-1. Wastewater Treatment Plants Selected for Analysis of Metal Speciation.**

| Site | Treatment Processes | Fe Addition? | Expected effluent pH | Expected [EDTA] |
|---|---|---|---|---|
| Grand Rapids | activated sludge | Yes | 7.3-7.8 | Moderate |
| SE San Francisco | pure oxygen-activated sludge | No | 6.0-6.5 | Moderate |
| San Jose/Santa Clara | activated sludge/BNR | No* | 7.3-7.8 | High |
| Baltimore | pure oxygen-activated sludge | Yes | 6.0-6.5 | Moderate |

Note:
*$FeCl_3$ is added to a major wastewater collection line for odor control.

Metal speciation was analyzed by combining results of several different analytical methods (Table 2-2). Combining results from techniques such as chelating resin column partitioning (CRCP) and direct metal-EDTA measurements made possible identification of any inconsistencies or artifacts in the measurements. For example, the concentration of strongly complexed Zn (ZnL), as determined by CRCP, should be greater than or equal to the concentration of $ZnEDTA^{2-}$. The analytical methods were sensitive enough to detect all of the Cu, Ni, and Zn species expected in wastewater, with the possible exception of $CuEDTA^{2-}$. $CuEDTA^{2-}$ was difficult to detect in wastewater because dissolved Cu often is present at concentrations close to or below the method detection limit for $CuEDTA^{2-}$ (70 nM or 4.4 μg/L).

**Table 2-2. Species Quantified in Samples Collected from Wastewater Treatment Plants.**

| Species | Abbreviation | Technique[1] | Detection limit[2] (nM) Total | Cu | Ni | Zn |
|---|---|---|---|---|---|---|
| Total EDTA | $EDTA_{tot}$ | IP HPLC | 20 | NA | NA | NA |
| FeEDTA complexes | $FeEDTA^-$ | IP HPLC | 20 | NA | NA | NA |
| Metal-EDTA complexes | $MeEDTA^{2-}$ | IX HPLC, IP HPLC | NA | 70 | 20 | 70 |
| Dissolved metals[3] | $Me_{diss}$ | GFAAS | NA | 15 | 15 | 1 |
| Particle-associated metals[4] | $Me_{part.}$ | GFAAS | NA | 15 | 15 | 1 |
| Labile metals | Me' | CRCP | NA | 1.5 | 1.5 | 0.1 |
| Complexed metals[5] | MeL | CRCP | NA | 1.5 | 1.5 | 0.1 |

Notes:

[1] IP HPLC: ion-pairing high performance liquid chromatography (Nowack et al., 1996); NA: not applicable; IX HPLC: ion-exchange high performance liquid chromatography with post-column reaction as described in Appendix A for quantification of $CuEDTA^{2-}$ and $ZnEDTA^{2-}$. IP HPLC as described by Bedsworth and Sedlak (1999) for $NiEDTA^{2-}$. GFAAS: Graphite furnace atomic absorption spectroscopy; CRCP: Chelating resin column chromatpgraphy (Sedlak et al., 1997).

[2] Detection limits for FeEDTA complexes, particle-associated metals and complexed metals are calculated by difference (particle-associated metals=total metals-dissolved metals). Detection limits depend on concentrations. Detection limits shown here assume relatively low concentrations of metals.

[3] Defined as those metals passing through a 0.45 □m filter.

[4] Determined by difference (total metals - dissolved metals).

[5] Determined by difference (dissolved metals - labile metals).

## 2.3.2 Metal Speciation in Wastewater

Concentrations of EDTA and pollutant metals in wastewater effluent samples varied among the treatment plants studied (Figure 2-1). At all four WWTPs, grab samples were collected to minimize changes in speciation during compositing or after sample collection. Samples were collected mid-morning on weekdays. To minimize contributions from stormwater infiltration, or, in the case of SESF WWTP, combined sewer inputs, most samples were collected during the dry season.

EDTA concentrations ranged from approximately 500 nM to 3,290 nM. The highest EDTA concentration was detected in the sample collected from the Grand Rapids WWTP. EDTA concentrations at the other WWTPs ranged from approximately 500 to 1,500 nM. EDTA concentrations at the SJSC WWTP in 1999 and in previous studies (Bedsworth and Sedlak, 1999) ranged from approximately 700 to 2,000 nM. EDTA concentrations in the four effluent samples collected from the SESF WWTP in 1999 ranged from 515 to 1,030 nM.

Among the four WWTPs, total concentrations of Cu, Ni, and Zn exhibited less variability than EDTA. Cu concentrations in all the wastewater effluent samples exceeded the most stringent discharge criterion (38 nM). Zn concentrations exceeded the most stringent criterion (890 nM) at all WWTPs except the SESF WWTP in June, and Ni only exceeded the most stringent criterion (140 nM) in the sample from the Grand Rapids WWTP. A significant relationship between total EDTA and total metals was not observed.

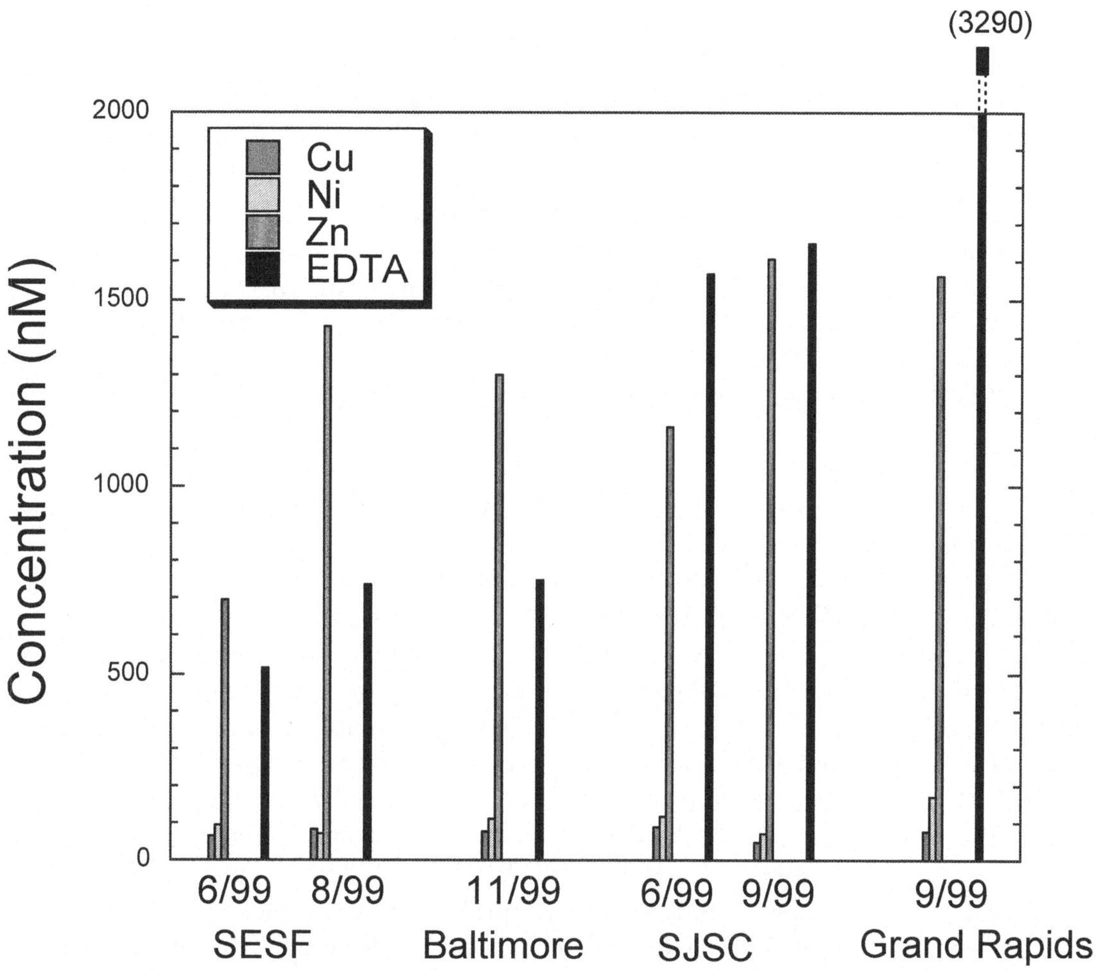


**Figure 2-1. Total Concentrations of Cu, Ni, Zn, and EDTA Measured in Wastewater Effluent.**

To obtain further insight into the role of EDTA in metal speciation in wastewater effluent, speciation data must be examined. To facilitate understanding the relationship between pH and speciation, results from the two oxygen-activated sludge WWTPs (Southeast San Francisco and Baltimore) and the two air-activated sludge WWTPs (San Jose/Santa Clara and Grand Rapids) are discussed together.

FeEDTA complexes accounted for most of the EDTA in the wastewater effluent at both oxygen-activated sludge WWTPs (compare the hollow and shaded EDTA bars for the three analyses depicted in Figure 2-2). Complexed forms of Zn (ZnL) accounted for a relatively small fraction of the total Zn (<10% at the SESF WWTP and <30% at the Baltimore WWTP). Most of the Ni in the effluent from both WWTPs was strongly complexed (NiL). Previous data (Bedsworth and Sedlak, 1999) suggested that the complexed Ni in wastewater effluent consists of $NiEDTA^{2-}$. However, matrix interference prevented analysis of $NiEDTA^{2-}$ at the SESF and Baltimore WWTPs. Most of the dissolved Cu in effluent samples from both facilities was strongly complexed (CuL). However, it was impossible to determine if the strong Cu complexes were $CuEDTA^{2-}$ because CuL was present at concentrations below the detection limit of the $CuEDTA^{2-}$ technique.

Results from the air-activated sludge WWTPs (two analyses at the SJSC WWTP and one at the Grand Rapids WWTP) indicated the presence of $CuEDTA^{2-}$, FeEDTA complexes, and $ZnEDTA^{2-}$ (Figure 2-3). FeEDTA complexes accounted for most of the EDTA at both WWTPs (approximately 70% at the SJSC WWTP and 85% at the Grand Rapids WWTP). The remainder of the EDTA consisted of $ZnEDTA^{2-}$ and $NiEDTA^{2-}$ (the sum of the three different metal-EDTA complexes was not statistically different from independently measured total EDTA). $ZnEDTA^{2-}$ accounted for approximately 90% and 65% of the strongly complexed Zn in the two samples from the SJSC WWTP and 50% of the strongly complexed Zn at the Grand Rapids WWTP, respectively. $NiEDTA^{2-}$ accounted for approximately 90% and 40% of the strongly complexed Ni at the SJSC WWTP and 75% of the strongly compelxed Ni at the Grand Rapids WWTP, respectively. Strongly complexed Cu at the SJSC WWTP was detected at concentrations near or below the detection limit of the $CuEDTA^{2-}$ technique. Strongly complexed Cu the Grand Rapids WWTP was extremely low (20 nM). $CuEDTA^{2-}$ was not detected in any of the samples.

Comparison of results from the oxygen- and air-activated sludge WWTPs suggested that pH plays an important role in metal speciation in wastewater effluents. These differences were most clearly evident for $ZnEDTA^{2-}$, which was detected at concentrations above 300 nM in the effluent of both air-activated sludge WWTPs but only in one effluent sample from an oxygen-activated sludge WWTP (at approximately 150 nM). When present in effluent from air-activated sludge WWTPs, $ZnEDTA^{2-}$ accounted for a significant percentage of dissolved Zn. The direct measurements of metal-EDTA complexes also were consistent with speciation measurements made by CRCP, because concentrations of metal-EDTA complexes were less than the measured concentrations of strongly complexed metals.

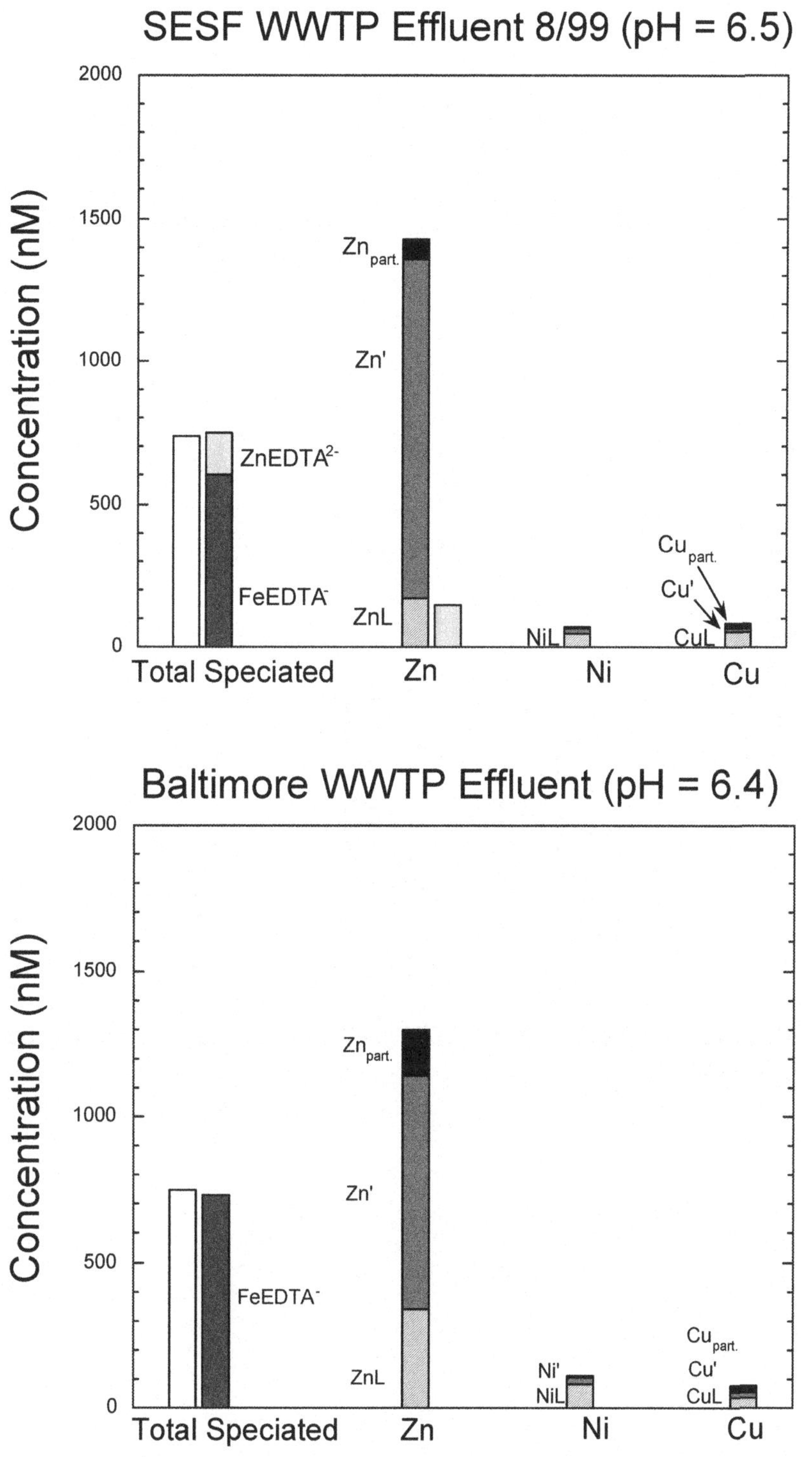


**Figure 2-2. Speciation of Cu, Ni, Zn, and EDTA in Effluent Samples Collected from Southeast San Francisco (SESF) and Baltimore Wastewater Treatment Plants.**

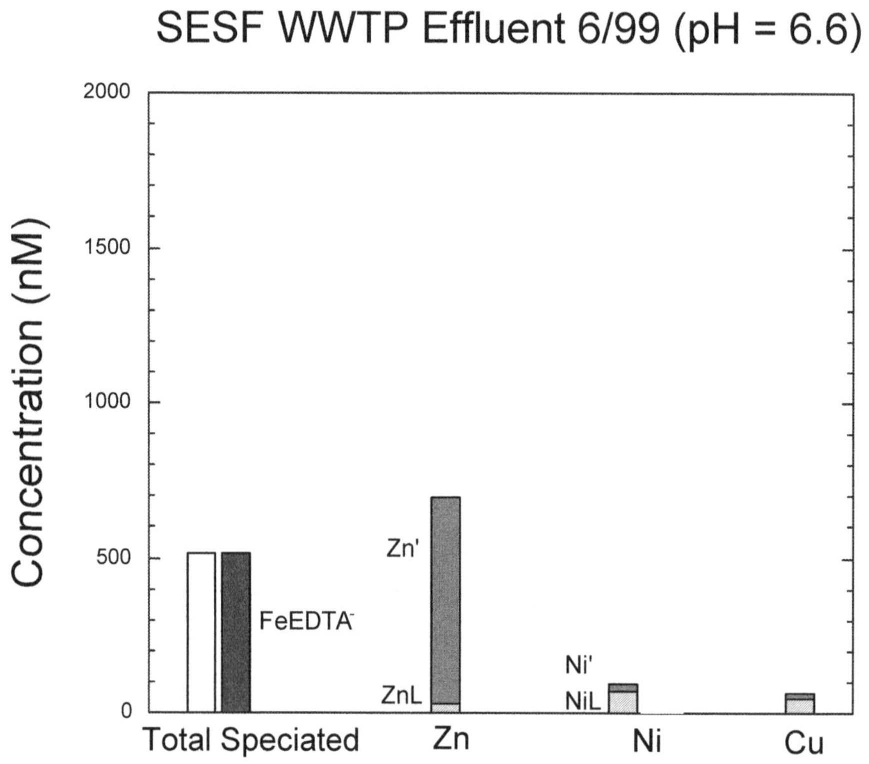


**Figure 2-2 (continued). Speciation of Cu, Ni, Zn, and EDTA in Effluent Samples Collected from Southeast San Francisco (SESF) Wastewater Treatment Plant.**

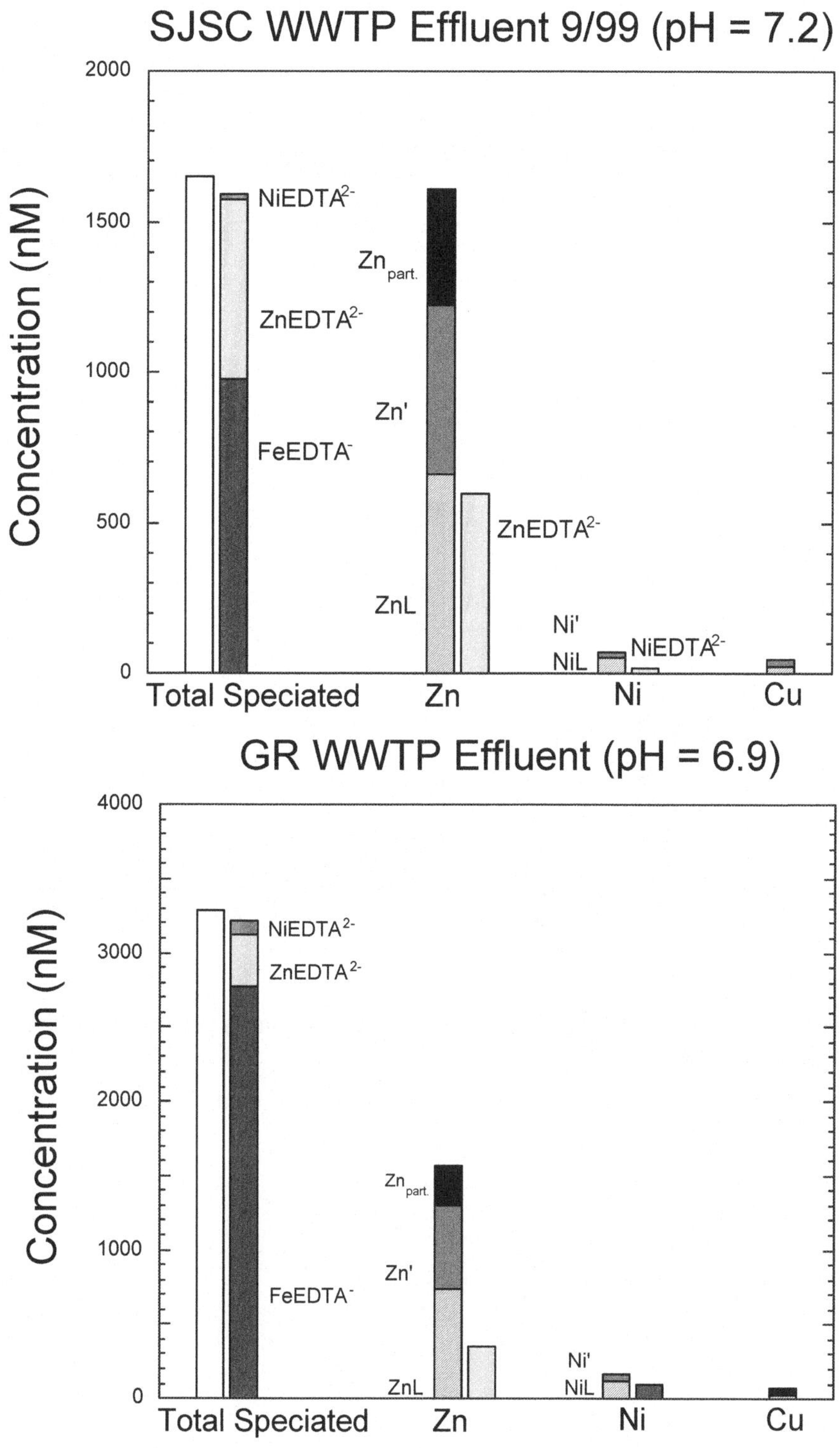


**Figure 2-3. Speciation of Cu, Ni, Zn, and EDTA in Effluent Samples Collected from San Jose/Santa Clara (SJSC) and Grand Rapids Wastewater Treatment Plants.**

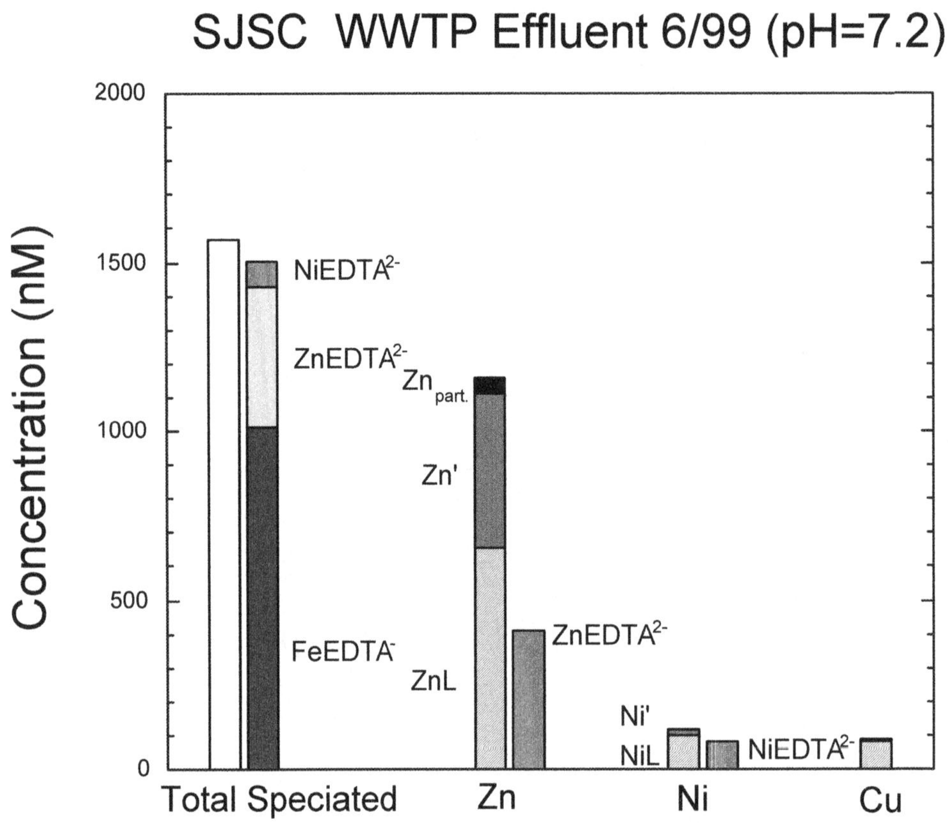


**Figure 2-3 (continued). Speciation of Cu, Ni, Zn, and EDTA in an Effluent Sample Collected from San Jose/Santa Clara (SJSC) Wastewater Treatment Plants (6/99).**

### 2.3.3 Changes in Metal Speciation during Wastewater Treatment

Measurements of metal speciation in wastewater effluent samples suggested that pollutant metal-EDTA complexes pass through air-activated sludge WWTPs but not oxygen-activated sludge WWTPs. To test the hypothesis that changes in EDTA speciation occur during oxygen-activated sludge treatment, the speciation of metals after primary treatment (referred to here as influent samples) at the four WWTPs discussed in the previous section was measured. Although grab samples collected within one to two hours of each other do not characterize the exact same parcel of water, they do provide insight into shifts in speciation that occur during treatment.

Results obtained at the SESF WWTP indicated that $CuEDTA^{2-}$ and $ZnEDTA^{2-}$ in the wastewater influent were converted into FeEDTA during treatment (Figure 2-4). This shift in speciation was consistent with predictions of equilibrium speciation calculations for the moderate pH drop that occurred between primary treatment (pH 7.3) and secondary treatment (pH 6.5). The pH actually dropped to a value below 6.5 in the covered oxygen-activated sludge basin and increased between the covered basin and the sampling point. After the $ZnEDTA^{2-}$ complexes dissociated, more than 90% of the Zn consisted of labile, dissolved complexes. Most of the Ni in the influent and effluent of the SESF WWTP was strongly complexed (NiL). Despite the matrix interference that prevented measuring $NiEDTA^{2-}$ at this site, data from other WWTPs (Bedsworth and Sedlak, 1999) suggest that the complexed Ni consists of $NiEDTA^{2-}$. The influent sample from the SESF WWTP was the only sample from the entire study in which $CuEDTA^{2-}$ was detected. $CuEDTA^{2-}$ accounted for more than 80% of the Cu in the influent. The high percentage of strongly complexed Cu in the effluent sample could have consisted of $CuEDTA^{2-}$; however, the concentration was below the method detection limit for the HPLC technique.

At the Baltimore WWTP, EDTA speciation did not change appreciably during treatment because EDTA in the plant's influent consisted almost entirely of FeEDTA complexes (Figure 2-5). One possible explanation for the speciation of EDTA in the influent is the relatively low pH (6.7), which could have been low enough to convert the metal-EDTA complexes into FeEDTA complexes prior to treatment. As a result of the influent EDTA speciation, changes in metal concentrations and metal speciation were attributable to phenomena unrelated to EDTA, such as metabolism of organic carbon and the effect of the small pH decrease on partitioning of metals onto particles. For example, solubilization of particle-associated Zn occurs during treatment, resulting in the production of labile and complexed Zn. Previous research (Rudd et al., 1984) indicated that Zn has a relatively low affinity for activated sludge. Although reasonably high concentrations of complexed Ni were detected in the effluent sample, little complexed or dissolved Ni was detected in the influent sample, and matrix interference prevented measurement of $NiEDTA^{2-}$ in either sample. Removal of complexed, labile, and particle-associated Cu also occurred at this WWTP.

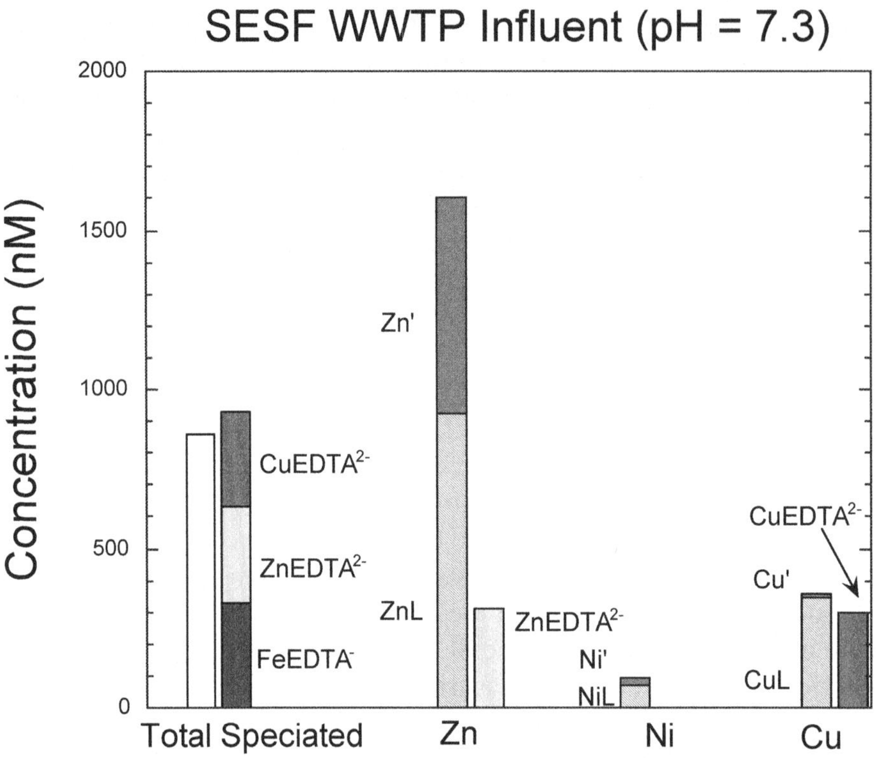


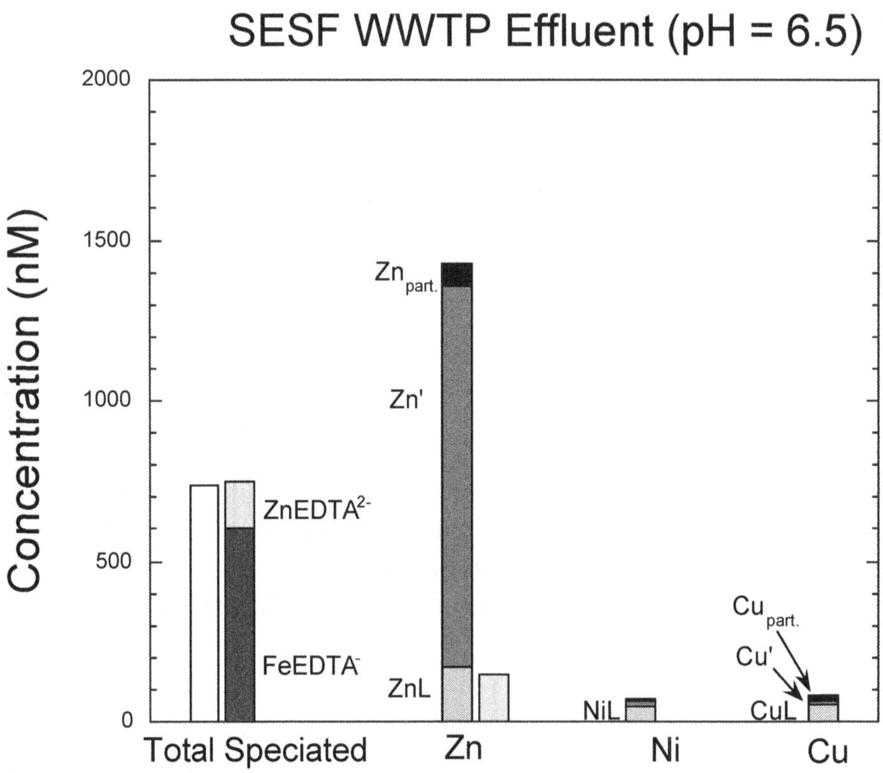


**Figure 2-4. Speciation of Cu, Ni, Zn, and EDTA in Influent and Effluent Samples Collected from Southeast San Francisco (SESF) Wastewater Treatment Plant, June 1999. Note: $NiEDTA^{2-}$ was not measured due to matrix interference. Particle-associated metals also were not measured.**

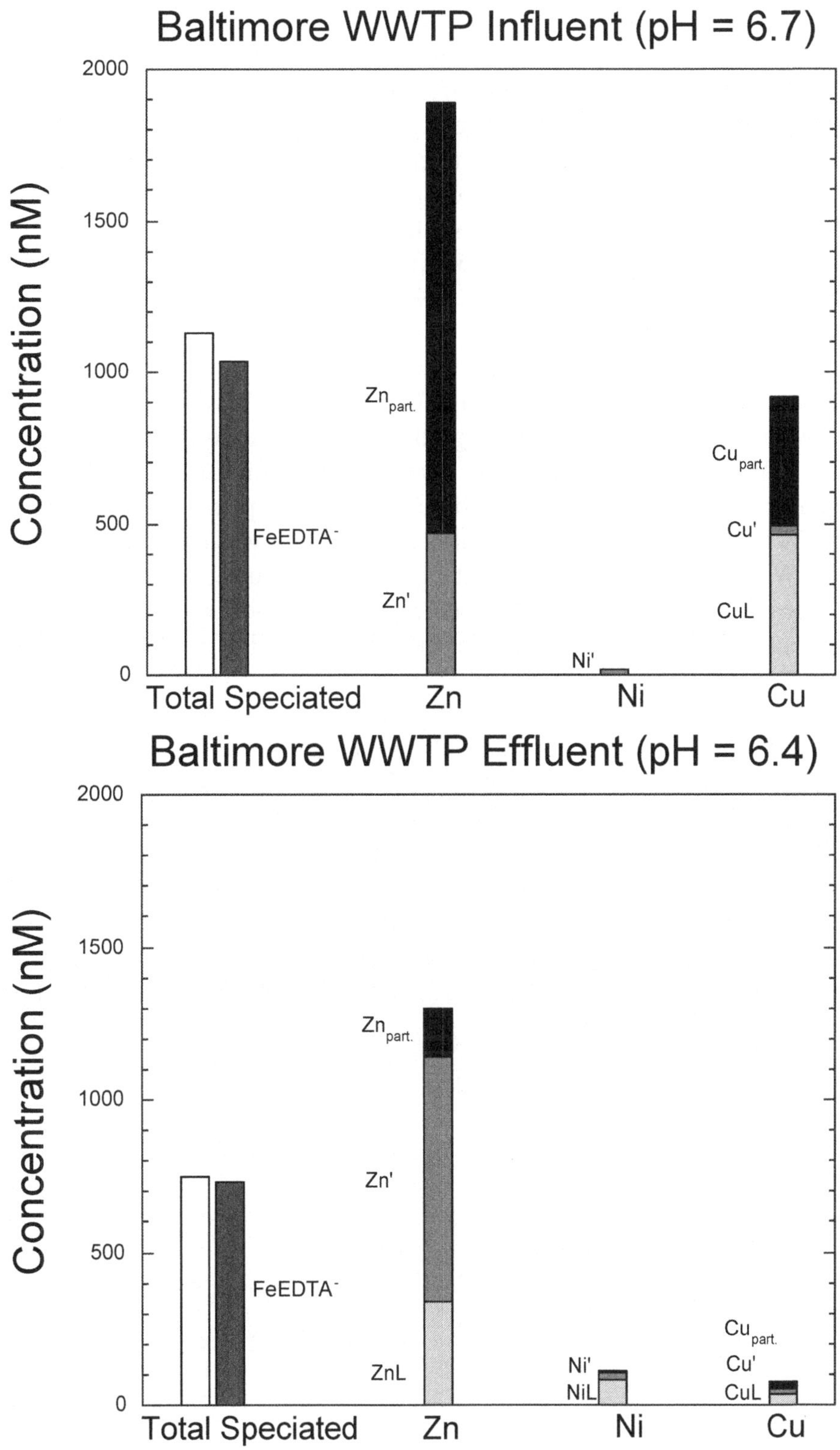


**Figure 2-5. Speciation of Cu, Ni, Zn, and EDTA in Influent and Effluent Samples Collected from Baltimore Wastewater Treatment Plant, November 4, 1999.**

At the SJSC WWTP (Figure 2-6), EDTA speciation did not change appreciably during treatment because the pH remained relatively constant (6.9 in influent vs. 7.2 in effluent). As observed at the Baltimore WWTP, production of labile Zn from particle-associated Zn occurred during secondary treatment. Treatment had a negligible effect on $ZnEDTA^{2-}$ and strongly complexed Ni. Although matrix interference prevented measurement of $NiEDTA^{2-}$ in the influent sample, the strong agreement between dissolved influent and dissolved effluent Ni (Ekster and Jenkins, 1996) and the consistent detection of $NiEDTA^{2-}$ in the plant's effluent suggest that much of the dissolved Ni entering the plant consists of $NiEDTA^{2-}$. The SJSC WWTP removed most of the particle-associated Cu (Ekster and Jenkins, 1996). Effluent Cu consists of a mixture of complexed and labile Cu species.

The speciation of metals and EDTA at the Grand Rapids WWTP also did not change significantly by treatment (Figure 2-7). Although the concentration of EDTA was much higher at the Grand Rapids WWTP than any of the other sites, the concentration of $ZnEDTA^{2-}$ and $NiEDTA^{2-}$ was comparable to that detected in other influent samples or in the effluent samples from the SJSC WWTP. Solubilization of particle-associated Zn was observed during treatment at the Grand Rapids WWTP, and overall, little net Zn removal occurred. As was the case at the SJSC WWTP, little removal of dissolved Ni occurred because most of it consisted of $NiEDTA^{2-}$. Cu removal was relatively low; however, unlike the other WWTPs, most of the Cu was associated with particles.

Comparison of data from the four WWTPs provided insight into the role of speciation in metal removal. Significant changes in EDTA speciation were only observed at the SESF WWTP, where the pH changed from approximately 7.0 prior to treatment to approximately 6.5 in the effluent. At the other three WWTPs, pH changes during treatment were too small to have a significant impact on speciation. The consistency of the overall effect of pH on EDTA speciation can be observed when all of the data are plotted together (Figure 2-8). Although these data were collected at facilities with different total EDTA and metal concentrations, they all were consistent with equilibrium calculations indicating that $ZnEDTA^{2-}$ and $CuEDTA^{2-}$ complexes will dissociate as the pH drops below 7.

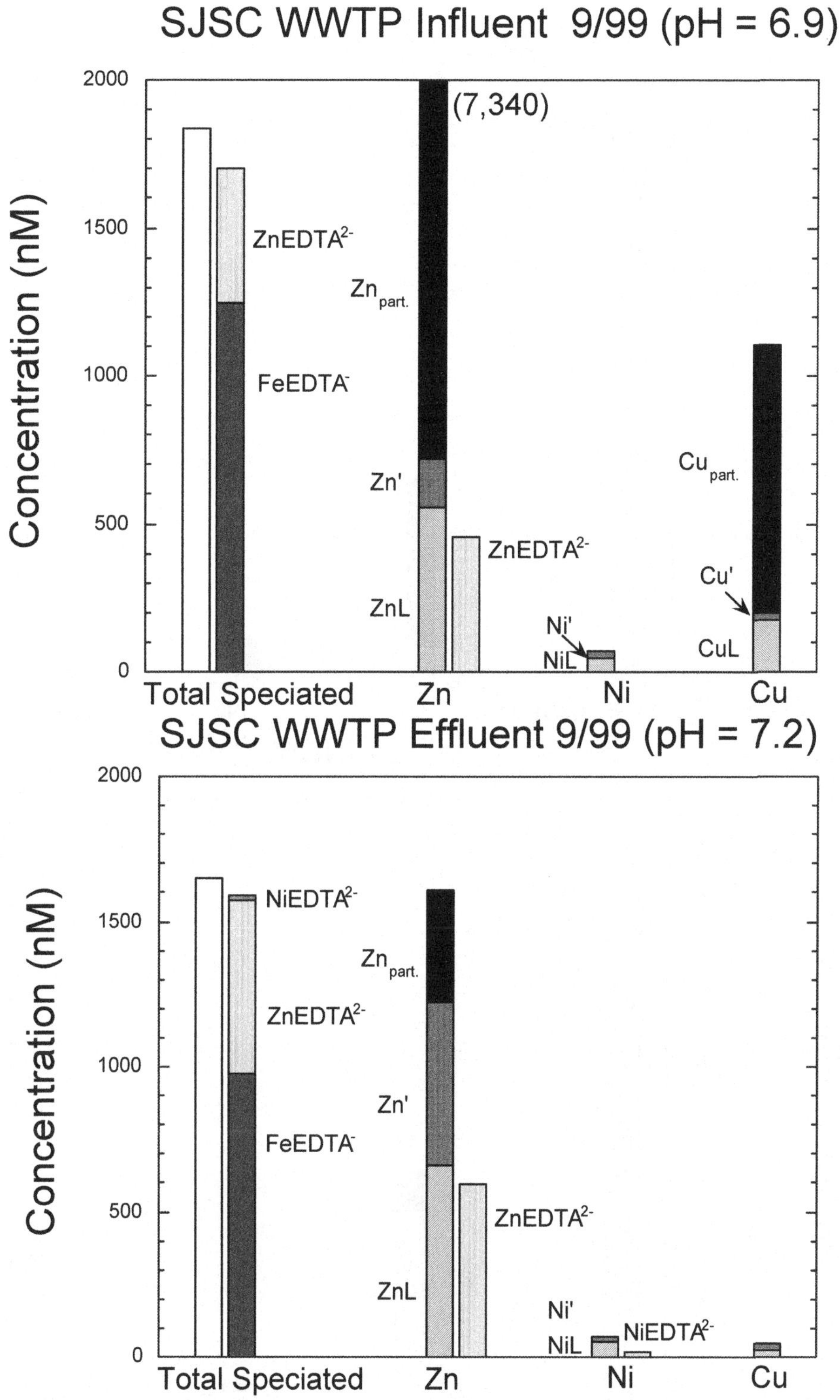


**Figure 2-6. Speciation of Cu, Ni, Zn, and EDTA in Influent and Effluent Samples Collected from San Jose/Santa Clara (SJSC) Wastewater Treatment Plant, September 1999.**

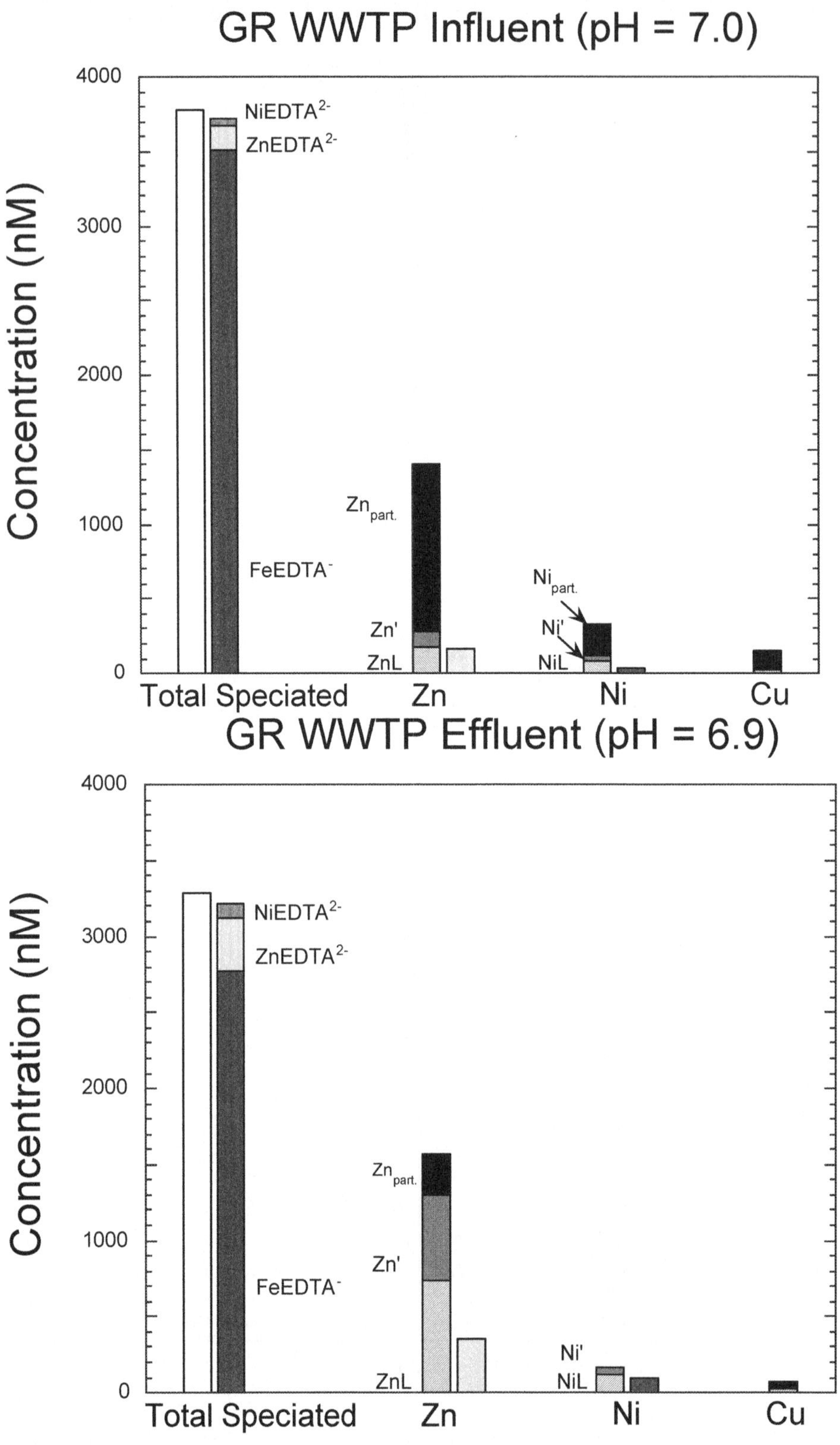


**Figure 2-7. Speciation of Cu, Ni, Zn, and EDTA in Influent and Effluent Samples Collected from Grand Rapids (GR) Wastewater Treatment Plant, September, 1999.**

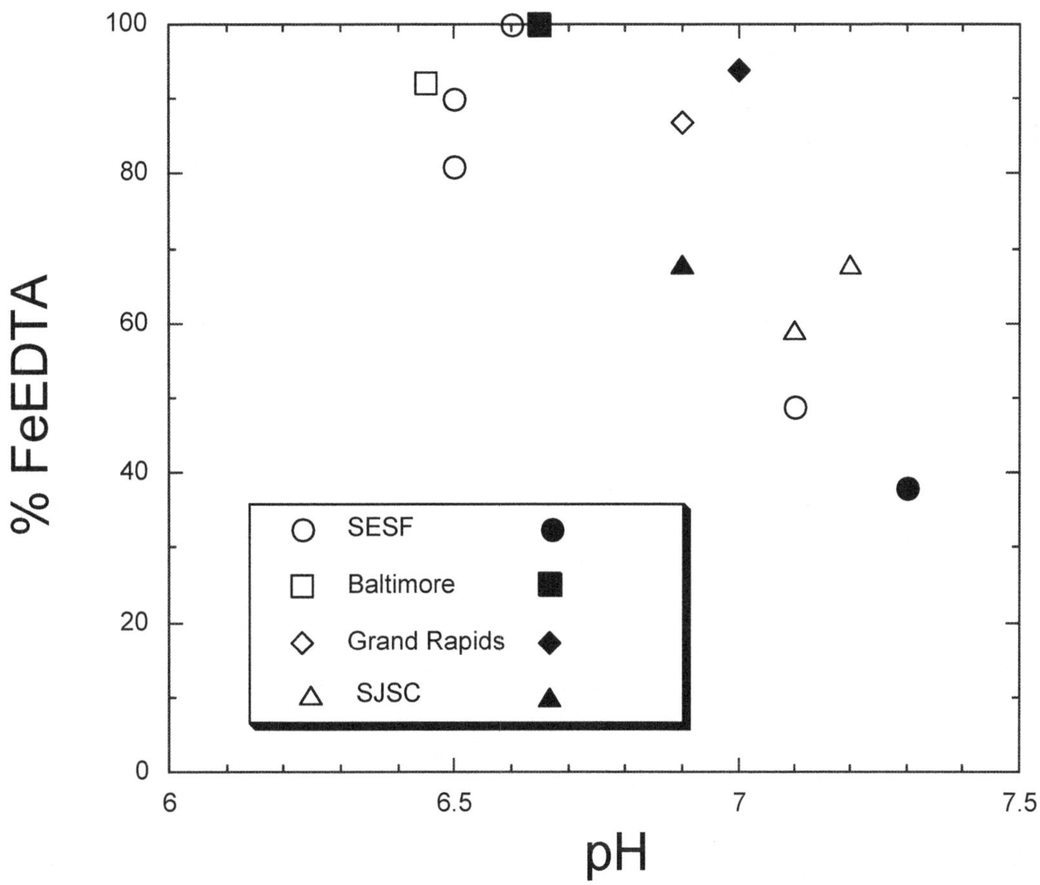


**Figure 2-8. Relationship Between pH and Percentage of FeEDTA Complexes in Influent (Filled Symbols) and Effluent (Hollow Symbols) Samples.**

CHAPTER 3.0

# Bench-Scale Studies to Assess the Effect of Chemical Addition During Primary Treatment on Metal Removal

## 3.1 Background

Results from the speciation measurements described in Chapter 2.0 indicated that metal-EDTA complexes play an important role in the removal of Cu, Ni, and Zn during municipal wastewater treatment. Furthermore, the data indicated a relationship between wastewater pH and EDTA speciation in which FeEDTA complexes accounted for a larger percentage of the EDTA as pH decreased below 7. This finding is important because it may be possible to exploit the conversion of pollutant metal-EDTA complexes into Fe EDTA complexes as a means of improving metals removal during wastewater treatment.

Any improvement in metals removal as the result of manipulation of EDTA speciation is expected to be relatively small because dissociation of the metal-EDTA complexes is unlikely to reach completion, and other processes also play a role in metal removal. However, a relatively small improvement in metal removal still would be useful, especially in locations where mass loading requirements are in place. For example, a 20% decrease in metals loading would allow flow to increase by 20% in a system where the metal mass loading is limiting the WWTP's expansion. Therefore, any experiments designed to evaluate the efficacy of treatment modifications must be capable of detecting small changes in metal concentrations. Because the composition of wastewater and treatment plant performance varies over time, the only practical means of detecting small changes in metal concentrations between different treatments involves the simultaneous analysis of treatments and controls. Therefore, candidate treatment processes were screened in bench-scale experiments.

Among the different options for manipulating conventional wastewater treatment processes to improve metal removal, the most attractive was chemical addition during primary treatment, which is commonly used to enhance particle removal; also the effects of Fe addition on the treatment process are well understood. Adding $FeCl_3$ during primary treatment was selected for evaluation because previous reports suggested that conversion of pollutant metal complexes to FeEDTA occurs after Fe(III) addition (Kari and Giger, 1996). The addition of HCl during primary treatment also was evaluated because the measurements of EDTA speciation in wastewater treatment plants described in Chapter 2.0 indicated that formation of FeEDTA complexes occurs when the pH is depressed in the pure oxygen/activated sludge systems. Furthermore, $FeCl_3$ stock solutions used in primary treatment are acidic and could depress solution pH. If pH depression improved metals removal, when the practice is implemented at full scale adding more acid to the stock solutions might be appropriate. However, a biological treatment system may not perform well when wastewater pH is depressed below approximately 6.

## 3.2 Bench-Scale Experiments

To test the hypothesis that manipulation of EDTA speciation can enhance metal removal in WWTPs, a bench-scale experiment was conducted using wastewater collected after the grit chamber at the Sacramento (Calif.) Regional WWTP (SR WWTP) and at the inlet to the primary settling chambers at the Millbrae (Calif.) WWTP. Immediately prior to initiation of the experiment, wastewater was collected in 10-L acid-cleaned HDPE containers and decanted into a series of HDPE containers. Depending upon the experiment, $FeCl_3$, from a 6.2 mM stock solution, or HCl, from a 0.2 mM stock solution, was added to the wastewater, which was allowed to settle to simulate primary treatment. After settling, an aliquot of the supernatant was mixed with an aliquot of return activated sludge (RAS) freshly collected from the plant. The samples were aerated with air delivered through a HDPE sparging bulb, followed by settling to simulate clarification. After settling, samples were collected from the upper layer of the container. The amount of time used for each of the settling and aeration steps varied slightly between experiments and as described below. Samples were analyzed for total and dissolved metals, metal speciation, total EDTA, and EDTA speciation as described in Chapter 2.0. Samples also were analyzed for pH, total suspended solids (TSS), and dissolved organic carbon (DOC).

### 3.2.1 Sacramento Regional WWTP

A total of five experiments were conducted at the SR WWTP. During the first, second and third experiments, preliminary data were collected, and the experimental design was evaluated. The bench-scale experiments were implemented during the fourth and fifth experiments. Overall, results of the bench-scale studies conducted at the SR WWTP indicate that chemical addition during primary treatment could result in a decrease of approximately 30% in effluent Zn concentrations. Addition of $FeCl_3$ with and without concurrent addition of HCl improved the removal of Cu and Zn during simulated primary treatment by reducing suspended solids and dissolved metals. A portion of the $ZnEDTA^{2-}$, which often accounted for a significant fraction of the effluent Zn, dissociated when chemicals were added during simulated primary treatment. Reducing suspended solids may not result in an improvement in metal removal because the solids normally are removed during secondary treatment. However, dissociation of $ZnEDTA^{2-}$ could result in a decrease in Zn concentrations after secondary treatment.

#### 3.2.1.1 Experiment 1

On April 26, 2000 a preliminary experiment was conducted at the SR WWTP to evaluate the conditions to be used in assessing the effect of chemical addition during primary treatment. The experiment was conducted using freshly collected, unfiltered wastewater, sampled after the grit chamber. To simulate primary settling, the samples were placed in 1-L polyethylene containers for one hour. To simulate secondary treatment, the supernatant from the primary settling step was decanted into 1-L polyethylene containers to which 210 mL of freshly collected RAS was added. Samples were collected after 4.8 hours.

The experiment subjected three treatments, one control, and one blank to simulated primary and secondary treatment:

- *pH adjustment:* The wastewater pH was adjusted to 6.0 using approximately 40 mL of 0.1 N HCl.
- $FeCl_3$ *addition:* A final $FeCl_3$ concentration of 50 mg/L was achieved by adding 50 mL of a 6.2 mM solution of $FeCl_3$.
- *pH adjustment and* $FeCl_3$ *addition:* $FeCl_3$ and HCl were added simultaneously for a final pH of 6.0 and a final $FeCl_3$ concentration of 50 mg/L.
- *Control:* A sample of wastewater without any treatment.
- *Blank:* A sample of deionized water.

The results are summarized in Table 3-1. The authors are cautious about interpretation of the data because the metal speciation was inadvertently perturbed prior to starting the experiment, when untreated wastewater was transported to the SR WWTP laboratory in sealed containers; unfiltered wastewater influent was collected in 1-L containers that were capped for approximately one hour prior to the experiment. The initial pH of the wastewater was 7.0. When the cap was removed and a sample was decanted for pH measurement immediately before starting the experiment, the pH was 6.8. Since the sample was aerated when it was decanted, the minimum pH in the bottle was likely less than 6.8. As indicated by previous studies, pH shifts in this region probably converted all the EDTA measured in the primary effluent into FeEDTA complexes. The absence of Cu and ZnEDTA complexes in all the primary effluent samples (including controls) was inconsistent with a previous measurement of wastewater influent at the SR WWTP, which indicated that only about 30% of the EDTA consisted of FeEDTA in the wastewater influent. Despite the absence of Cu and ZnEDTA complexes in the wastewater, some potentially useful observations can be made about the effects of pH reduction and/or $FeCl_3$ addition on metal removal.

**Table 3-1. Results of Samples Collected after Simulated Primary Treatment for Chemical Addition to Primary Treatment Experiment Conducted at SR WWTP on April 26, 2000 (Experiment 1).**

| | Concentration (nM) | | | | | |
|---|---|---|---|---|---|---|
| | After Primary Treatment | | | After Secondary Treatment | | |
| Treatment | $[Cu]_{diss}$ | $[Ni]_{diss}$ | $[Zn]_{diss}$ | $[Cu]_{diss}$ | $[Ni]_{diss}$ | $[Zn]_{diss}$ |
| Control | 120 | 55 | 880 | 58 | <35 | 430 |
| HCl | 80 | <35 | 820 | 38 | 49 | 420 |
| $FeCl_3$ | 110,88 | 72,97 | 540,640 | 58 | <35 | 380 |
| HCl & $FeCl_3$ | 58 | 85 | 590 | 42,46 | 49,80 | 430,370 |

| | After Primary Treatment | | | | After Secondary Treatment | | | |
|---|---|---|---|---|---|---|---|---|
| Treatment | TSS (mg/L) | pH after primary | $[EDTA]_{total}$ (nM) | $[FeEDTA]_{total}$ (nM) | TSS (mg/L) | pH after secondary | $[EDTA]_{total}$ (nM) | $[FeEDTA]_{total}$ (nM) |
| Control | 51 | 6.9 | 850 | 850 | 88 | 7.0 | 900 | 810 |
| HCl | 57 | 6.0 | 910 | 910 | 92 | 6.5 | 700 | 670 |
| $FeCl_3$ | 58,59 | 6.6 | 1070,1010 | 1070,1010 | 88 | 7.0 | 900 | 810 |
| HCl & $FeCl_3$ | 68 | 5.8 | 1120 | 1120 | 130,120 | 6.6 | 830,850 | 780,470 |

Note: Results of duplicate treatments are separated with commas.

The three treatments each affected concentrations of dissolved metals present in primary effluent relative to controls (Figure 3-1). For Zn, reduction of pH had little or no effect on concentrations. However, adding $FeCl_3$, with or without concurrent addition of acid, reduced dissolved Zn by approximately 30%. For Cu, both pH reduction and $FeCl_3$ addition lowered dissolved concentrations by approximately 30%. Concurrent addition of $FeCl_3$ and acid reduced dissolved Cu by approximately 50%. For Ni, adding $FeCl_3$ or acid and $FeCl_3$ increased concentrations by over 30%.

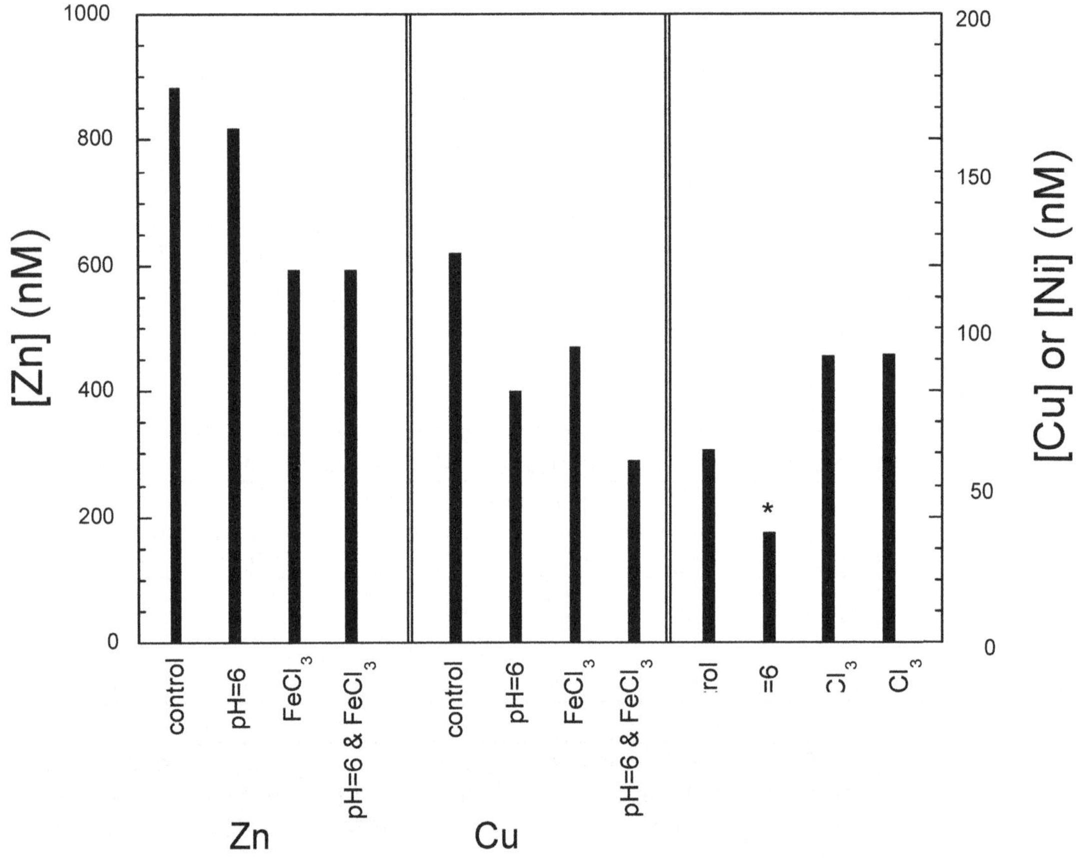


**Figure 3-1. Concentrations of Dissolved Zn, Cu, and Ni Measured after Simulated Primary Treatment on April 26, 2000 (Experiment 1). Note that Zn data are on a different scale than Cu and Ni data. * Indicates concentration below limit of quantification, datum plotted at detection limit.**

In simulated secondary effluent, the treatments affected concentrations of Cu and Ni, but not Zn, relative to control (Figure 3-2). For Cu, pH reduction with or without $FeCl_3$ addition resulted in a 35% decrease in dissolved concentrations, while addition of $FeCl_3$ alone had no effect. For Ni, pH reduction with or without $FeCl_3$ addition increased dissolved Ni by approximately 35%, while addition of $FeCl_3$ alone had no effect.

Although it is difficult to interpret the limited number of observations, some of the findings are consistent with expectations. For example, the observation that $FeCl_3$ addition lowered the concentrations of Zn and Cu but not Ni during primary treatment is consistent with observations that Ni has a relatively low affinity for hydrous ferric oxide ($FeOOH_{(s)}$). The observation that dissolved Ni increased when pH was lowered was consistent with previous data on affinity of Ni for surfaces. Other aspects of the data were inconclusive and the results of this experiment were considered preliminary.

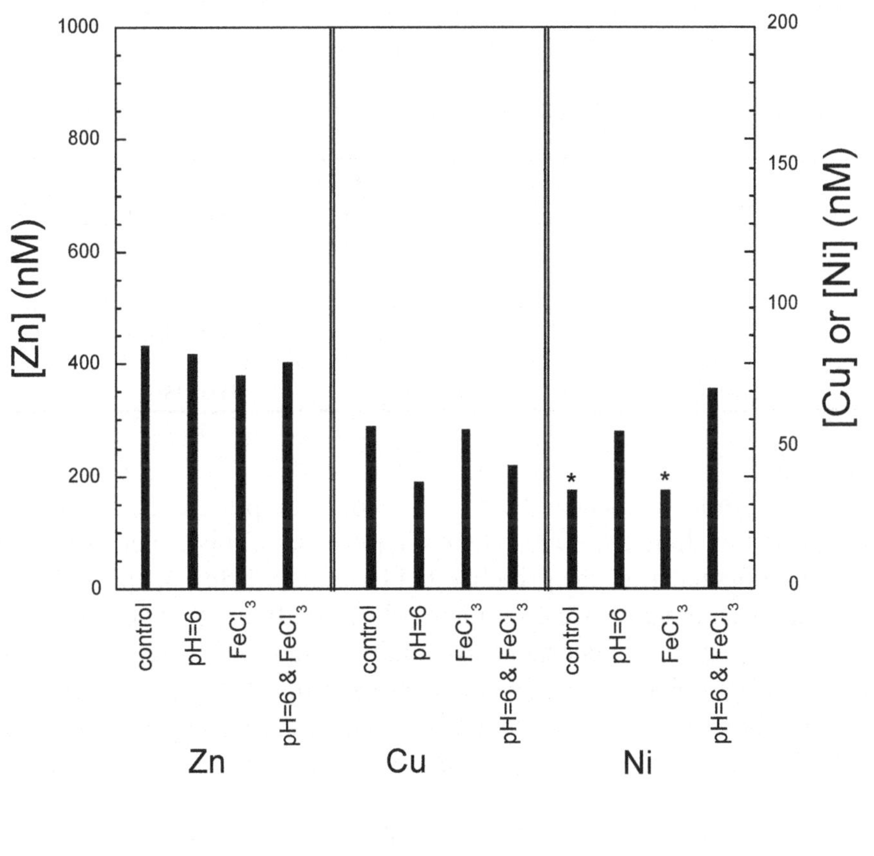


**Figure 3-2. Concentrations of Dissolved Zn, Cu, and Ni Measured after Simulated Secondary Treatment on April 26, 2000 (Experiment 1). Note that Zn data are on a different scale than Cu and Ni data. * Indicates concentration below limit of quantification, datum plotted at detection limit.**

### 3.2.1.2 Experiment 2

On May 30, 2000, another experiment was conducted at the SR WWTP after modifying the experimental protocol to address some of the difficulties observed during the first experiment. In this experiment, attention was focused on the role of pH adjustment in metal speciation. In particular, samples were collected in 2-L bottles to minimize pH changes resulting from oversaturation of the samples with carbon dioxide. To verify the effect of pH on EDTA speciation, results from pH adjustment experiments were compared with measured speciation of EDTA at the full-scale treatment plant (Table 3-2).

The changes in EDTA speciation measured at full scale were consistent with previous observations at the SR WWTP and the other WWTPs analyzed previously: the percentage of FeEDTA complexes increased from 28% in the influent to 54% in the effluent as the pH dropped. The results of the bench-scale experiments were ambiguous because the duplicate control experiments yielded inconsistent results. In one sample, the EDTA speciation was consistent with measurements from primary effluent (35% of the EDTA consisted of FeEDTA complexes), while the other sample exhibited a shift in EDTA speciation comparable to that observed when the pH was depressed during secondary treatment (62% of the EDTA consisted of FeEDTA complexes). Samples subjected to pH adjustment exhibited the expected formation of FeEDTA complexes, but not to the extent observed in previous measurements at other WWTPs. As a result of these analytical problems, data interpretation was not possible.

**Table 3-2. Concentrations of EDTA Complexes at SR WWTP on May 30, 2000 (Experiment 2).**

| | Concentration (nM) | | | |
|---|---|---|---|---|
| **Sample** | **$[EDTA]_{total}$** | **$[NiEDTA^{2-}]$** | **$[FeEDTA]_{total}$** | **% FeEDTA** |
| 1° Effluent | 740 | 79 | 210 | 28 |
| 2° Effluent | 670 | 82 | 360 | 54 |
| | chemical addition to wastewater collected after grit chamber | | | |
| Control | 320, 400 | <50, 100 | 200, 140 | 62, 35 |
| pH 6.3 | 422 | 98 | 220 | 52 |
| pH 6.0 | 400, 490 | 79, 160 | 180, 250 | 45, 51 |

### 3.2.1.3 Experiment 3

To resolve uncertainties associated with the second sampling event, a third round of sampling was conducted on June 29, 2000. The main purpose of this experiment was to determine if samples collected after the grit chamber exhibited different speciation from samples collected after primary treatment or if the samples collected after the grit chamber behaved differently during pH adjustment experiments. In addition, samples were collected to determine if syringe-filtered samples yielded the same results as cartridge-filtered samples. (In previous experiments, syringe filters were used for samples collected after the grit chamber, while cartridge filters were used for samples collected after primary and secondary treatment.)

Results of the experiment indicated that samples collected after the grit chamber and after primary treatment exhibited identical behavior and that both types of filters yielded comparable results (Table 3-3). No differences in total EDTA concentrations or percentage EDTA was observed for samples collected after the grit chamber and after primary treatment. In addition, acidification of both types of samples resulted in conversion of approximately the same percentage of the EDTA into FeEDTA complexes.

Table 3-3. Concentrations of EDTA Complexes at SR WWTP on June 29, 2000 (Experiment 3).

| Filtration Method | pH | $[EDTA]_{total}$ | $[FeEDTA]_{total}$ | % FeEDTA |
|---|---|---|---|---|
| | | Samples collected after grit chamber | | |
| cartridge | 6.7 | 690 | 560 | 81 |
| | | 610 | 470 | 77 |
| | | 620 | 480 | 77 |
| syringe | 6.7 | 660,690 | 580,570 | 88,83 |
| | | 690 | 570 | 83 |
| | | 620 | 530 | 86 |
| | 5.6 | 600 | 529 | 88 |
| | 5.2 | 620 | 570 | 92 |
| | | Samples collected after primary treatment | | |
| cartridge | 6.7 | 670,680 | 520,610 | 78,90 |
| | | 700 | 580 | 83 |
| | | 680 | 560 | 82 |
| syringe | 6.7 | 680,670 | 530,600 | 78,90 |
| | | 660 | 510 | 77 |
| | | 700 | 560 | 80 |
| | 5.5 | 730 | 680 | 94 |
| | 5.4 | 670,710 | 620,670 | 93,95 |

A more significant finding of the third sampling trip was that most of the EDTA entering the SR WWTP consisted of FeEDTA complexes. The speciation of EDTA in the plant's influent is consistent with the trend observed at the WWTPs sampled during the second year of the project. In a preliminary sample in 1999 and the sample collected in conjunction with Experiment 2, FeEDTA complexes accounted for less than 50% of the EDTA. However, FeEDTA accounted for approximately 80% of the EDTA in samples collected in association with Experiment 3. These results suggest that the inconsistencies in influent EDTA speciation could have been attributable to variations in influent pH.

To further assess the possibility that EDTA speciation would vary at the SR WWTP, historic data was obtained on influent pH at the SR WWTP (Figure 3-3). Data from 1998 indicated that mean influent pH values ranged from 6.7 to 6.9. Therefore, on the basis of equilibrium predictions and data from other WWTPs (Figure 2-8), the influent speciation of EDTA was expected to vary, with FeEDTA accounting for anywhere from approximately 20% to -90% of the total EDTA entering the plant.

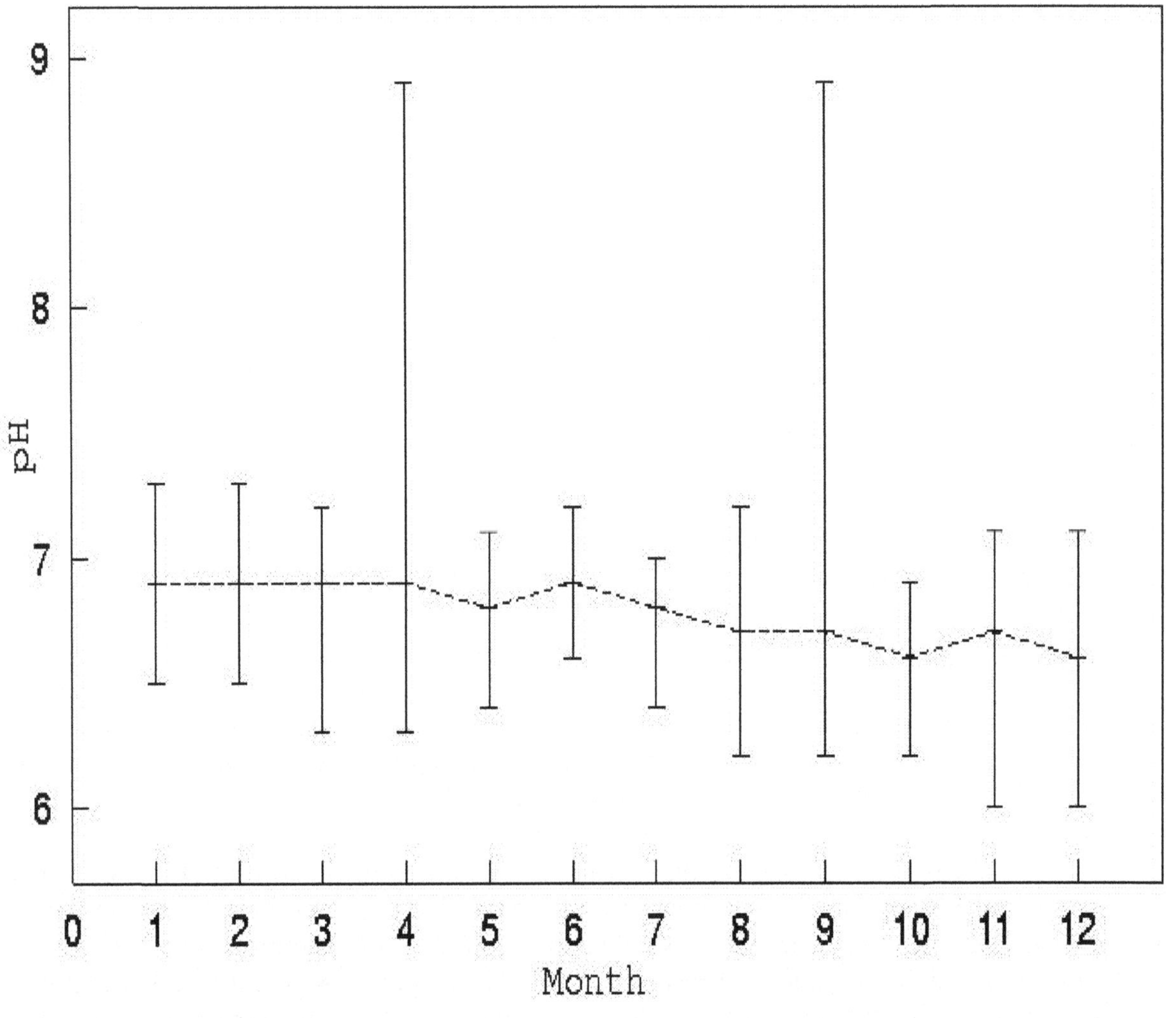


**Figure 3-3. Influent pH at SR WWTP in 1998. Lines Represent Rrange and Average for pH Values Observed in Daily Grab Samples.**

### 3.2.1.4 Experiment 4

After completing the second and third sampling trips to the SR WWTP, the authors designed an experimental protocol for the bench-scale experiments that would meet project objectives. On July 24, 2000, a fourth experiment was performed on the effect of chemical addition on metal removal. Results of the experiment (summarized in Figures 3-4, 3-5 and Tables 3-4, 3-5) suggested that chemical addition during primary treatment had the potential to improve the removal of Cu and Zn. Results observed after simulated primary and after secondary treatment are discussed below.

Addition of $FeCl_3$, with or without the concurrent addition of HCl, reduced the total concentrations of Cu and Zn by approximately 30%, relative to controls, in samples from the simulated primary treatment step (Figure 3-4). Addition of HCl alone increased the total concentrations of Cu and Zn by 10% to 20%. For the treatment with $FeCl_3$ alone, the decrease in metals concentrations was attributable almost entirely to the removal of dissolved metals: dissolved metal concentrations decreased by more than 95% relative to control after $FeCl_3$ treatment. The increase in total Zn after addition of HCl was attributable to release of Zn from the particles. This finding was consistent with expectations that Zn would have a lower affinity for particles when the pH was reduced. The increase in total Cu concentration after HCl addition appeared to be attributable to an increase in the concentration of particle-associated Cu, which was consistent with TSS data (Table 3-4). After treatment with both HCl and $FeCl_3$, dissolved metals concentrations increased relative to additional of $FeCl_3$ alone. Overall metals removal for the combined treatment was nearly identical to treatment with only $FeCl_3$.

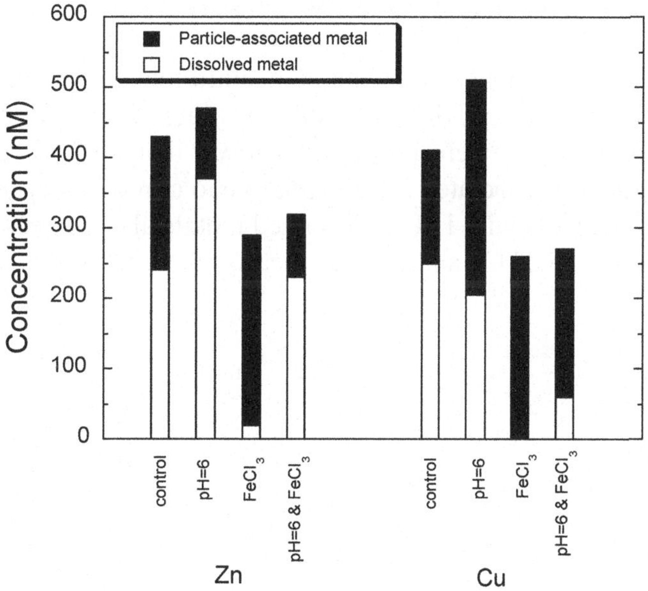


**Figure 3-4. Concentrations of Metals Measured at SR WWTP after Simulated Primary Treatment on July 24, 2000 (Experiment 4).**

**Table 3-4. Results of Samples Collected after Simulated Primary Treatment for Chemical Addition to Primary Treatment Experiment Conducted at SR WWTP on July 24, 2000 (Experiment 4).**

| | Concentration (nM) | | | | | | | | |
|---|---|---|---|---|---|---|---|---|---|
| **Treatment** | $[Cu]_{tot}$ | $[Cu]_{diss}$ | [CuL] | $[Ni]_{tot}$ | $[Ni]_{diss}$ | [NiL] | $[Zn]_{tot}$ | $[Zn]_{diss}$ | [ZnL] |
| Control | 410 | 250 | 240 | <50 | <50 | NA | 430 | 240 | 110 |
| HCl | 510 | 190,220 | 180 | <50 | <50 | NA | 470 | 330,410 | 130 |
| $FeCl_3$ | 260 | 2 | 2 | <50 | <50 | NA | 290 | <20 | <20 |
| HCl & $FeCl_3$ | 270 | 60 | 50 | <50 | <50 | NA | 320 | 230 | NA |

| Treatment | TSS (mg/L) | pH | $[EDTA]_{total}$ (nM) | $[FeEDTA]_{total}$ (nM) | % FeEDTA |
|---|---|---|---|---|---|
| Control | 74,60 | 6.7 | 560 | 350 | 63 |
| HCl | 90 | 6.0,6.0 | 550,540 | 360,320 | 62 |
| $FeCl_3$ | 60 | 6.6 | 530 | 330 | 62 |
| HCl & $FeCl_3$ | 78 | 5.6 | 480 | 400 | 83 |

NA = analysis not available.

Analysis of EDTA speciation indicated that FeEDTA complexes accounted for 60% in the wastewater prior to treatment. After simulated primary treatment, total EDTA concentrations ranged from 480 to 560 nM, and the percentage of FeEDTA complexes was unchanged for all treatments except the addition of both $FeCl_3$ and HCl, in which FeEDTA complexes increased to 83% of the total EDTA. Neither $CuEDTA^{2-}$ nor $ZnEDTA^{2-}$ were detected by the direct measurement technique. It seems unlikely that the pollutant metal-EDTA complexes were present at concentrations just below their detection limits (75 nM) because $CuEDTA^{2-}$ or $ZnEDTA^{2-}$ would not be expected to be removed by adding $FeCl_3$. Therefore, EDTA had a relatively small effect on metal removal during primary treatment.

After simulated secondary treatment, concentrations of Zn and Cu were considerably lower than those measured after primary treatment (Figure 3-5). All three treatments yielded total Zn and Cu concentrations that were 15% to 35% lower than control. The decrease in concentrations of particle-associated metals, relative to control, was consistent with the TSS data. For the two treatments with $FeCl_3$, dissolved metals also decreased by approximately 30% relative to control. Dissolved Cu and Zn were nearly identical to the control for the treatment with HCl alone.

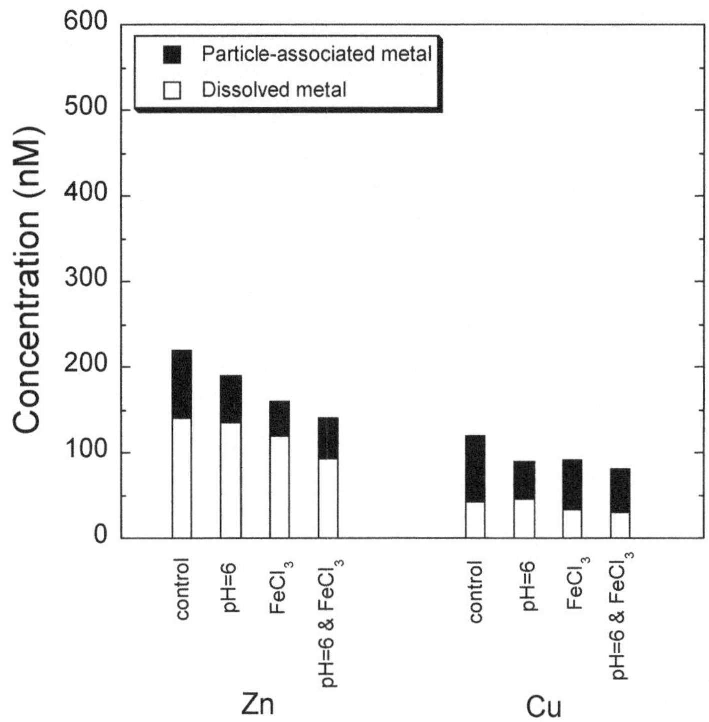


**Figure 3-5. Concentrations of Metals Measured at SR WWTP after Simulated Secondary Treatment on July 24, 2000 (Experiment 4).**

**Table 3-5. Results of Samples Collected after Simulated Secondary Treatment for Chemical Addition to Primary Treatment Experiment Conducted at SR WWTP on July 24, 2000 (Experiment 4).**

| | Concentration (nM) | | | | | | | | |
|---|---|---|---|---|---|---|---|---|---|
| **Treatment** | $[Cu]_{tot}$ | $[Cu]_{diss}$ | [CuL] | $[Ni]_{tot}$ | $[Ni]_{diss}$ | [NiL] | $[Zn]_{tot}$ | $[Zn]_{diss}$ | [ZnL] |
| Control | 120 | 43 | 40 | <50 | <50 | NA | 220 | 140 | 150 |
| HCl | 90 | 50,41 | 48,38 | <50 | <50 | NA | 190 | 130,140 | 200 |
| $FeCl_3$ | 91 | 33 | 26 | <50 | <50 | NA | 160 | 120 | NA |
| HCl & $FeCl_3$ | 81 | 30 | 28 | <50 | <50 | NA | 140 | 93 | NA |

| Treatment | TSS (mg/L) | $[EDTA]_{total}$ (nM) | $[FeEDTA]_{total}$ (nM) | *% FeEDTA* |
|---|---|---|---|---|
| Control | 15 | 590 | 360 | 61 |
| HCl | 7 | 570,550 | 430,410 | 75 |
| $FeCl_3$ | 7 | 530 | 400 | 75 |
| HCl & $FeCl_3$ | 9 | 580 | 490 | 84 |

NA = analysis not available.

After simulated secondary treatment, the total concentration of EDTA remained constant for all treatments, and the percentage of FeEDTA complexes remained constant for the control and treatment with both HCl and $FeCl_3$. The percentage of FeEDTA complexes increased from approximately 60% to 75% for the treatments with HCl or $FeCl_3$ relative to samples collected after primary treatment.

Overall, the results from experiment 4 led the authors to reexamine the merits of testing the hypotheses at the SR WWTP. Although chemical addition appeared to have an effect on the speciation of EDTA, the differences between the treatments and controls were difficult to measure because FeEDTA accounted for 60% of the initial EDTA. Despite the unsuitable nature of the SR WWTP for testing hypotheses about the role of EDTA in the treatment process, results of experiment 4 suggested that the removal of Cu and Zn could be enhanced by chemical addition during primary treatment. In particular, adding $FeCl_3$, which is a relatively common treatment chemical, appeared to reduce metals concentrations after both primary and secondary treatment. The effect appeared to be attributable to a combination of enhanced particle removal during primary treatment and dissociation of approximately half the $ZnEDTA^{2-}$. The actual effect of chemical addition to primary would have been greater in a treatment plant because the RAS used in these experiments was collected from a treatment plant where $FeCl_3$ addition was not in use. If $FeCl_3$ were added during primary treatment at the SR WWTP, it is likely that the concentration of metals in the RAS would be lower than the RAS used for these experiments. Therefore, a lower mass of metals would be contributed by the RAS.

**3.2.1.5 Experiment 5**

To verify the effect of $FeCl_3$ addition on the removal of metals at the SR WWTP, a fifth experiment was conducted on September 20, 2000. Results from this experiment confirmed the previous observations that adding $FeCl_3$ improved the removal of Cu and Zn during primary treatment (Figure 3-6, Table 3-6). Lower concentrations of dissolved and particle-associated metals were observed for all three treatments, relative to control. The decrease in particle-associated metals was consistent with the decrease in total suspended solids that occurred when $FeCl_3$ was added. The decrease in dissolved metal concentrations was consistent with measurements of metal speciation. In this experiment, direct measurements of $ZnEDTA^{2-}$ were unreliable due to problems encountered in post-column conversion of the complexes. However, measurements of nonFeEDTA complexes, which were presumed to consist almost entirely of $ZnEDTA^{2-}$, indicated that $ZnEDTA^{2-}$ was partially removed during treatment. Measurements of dissolved Cu indicated that almost all was complexed and that complexed Cu was removed after $FeCl_3$ addition. At the highest $FeCl_3$ dose, approximately half of the complexed Cu was removed.

The overall effect of $FeCl_3$ addition on metal removal after secondary treatment was small (Figure 3-7, Table 3-7). Differences in particle-associated metal concentrations among treatments were mostly attributable to differences in total suspended solids. In this experiment, total suspended solids concentrations were elevated in the treatments with 30 and 45 mg/L of $FeCl_3$, relative to control. However, these differences were attributed to the decantation process used in these experiments rather than any effect of the treatment. A decrease in dissolved Zn and nonFeEDTA complexes, but not dissolved Cu, was also observed relative to control. These results suggested that addition of $FeCl_3$ reduced effluent Zn concentrations by approximately 30%. Because little $CuEDTA^{2-}$ normally is present, addition of $FeCl_3$ is likely to have a much smaller effect on Cu removal.

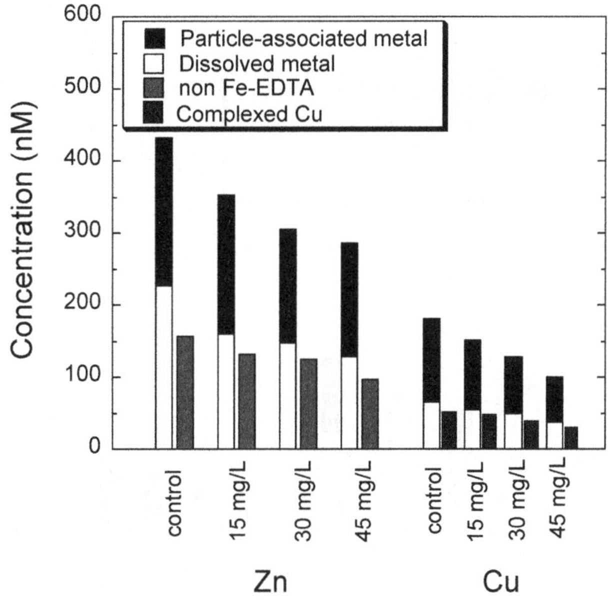


**Figure 3-6. Speciation of Cu and Zn after Simulated Primary Treatment in Samples Collected from SR WWTP on September 20, 2000 (Experiment 5).**

**Table 3-6. Results of Samples Collected after Simulated Primary Treatment for Chemical Addition to Primary Treatment Experiment Conducted at SR WWTP on September 20, 2000 (Experiment 5).**

| | Concentration (nM) | | | | | |
|---|---|---|---|---|---|---|
| **Treatment** | $[Cu]_{tot}$ | $[Cu]_{diss}$ | [CuL] | $[Zn]_{tot}$ | $[Zn]_{diss}$ | [ZnL] |
| Control | 180 | 64 | 54 | 430 | 230 | NA |
| 15 mg/L $FeCl_3$ | 150 | 55 | 45 | 360 | 160 | NA |
| 30 mg/L $FeCl_3$ | 130 | 49 | 41 | 310 | 150 | NA |
| 45 mg/L $FeCl_3$ | 100 | 35,41 | 27 | 280 | 130 | NA |

| **Treatment** | **TSS (mg/L)** | **pH before** | **pH after** | $[EDTA]_{total}$ (nM) | $[FeEDTA]_{total}$ (nM) | **% FeEDTA** |
|---|---|---|---|---|---|---|
| Control | 98 | 6.85 | 7.07 | 610 | 460 | 75 |
| 15 mg/L $FeCl_3$ | 80 | 6.87 | 6.92 | 560 | 430 | 77 |
| 30 mg/L $FeCl_3$ | 62 | 6.88 | 6.79 | 600 | 480 | 80 |
| 45 mg/L $FeCl_3$ | 60 | 6.87 | 6.65 | 520 | 420 | 81 |

NA = analysis not available.

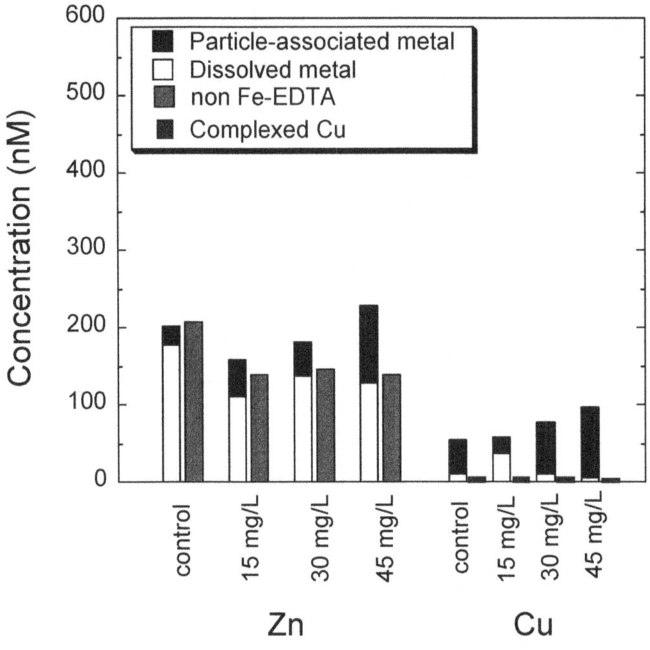


**Figure 3-7. Speciation of Cu and Zn after Simulated Secondary Treatment in Samples Collected from SR WWTP on September 20, 2000 (Experiment 5).**

**Table 3-7. Results of Samples Collected after Simulated Secondary Treatment for Chemical Addition to Primary Treatment Experiment Conducted at SR WWTP on September 20, 2000 (Experiment 5).**

| | Concentration (nM) | | | | | |
|---|---|---|---|---|---|---|
| **Treatment** | $[Cu]_{tot}$ | $[Cu]_{diss}$ | [CuL] | $[Zn]_{tot}$ | $[Zn]_{diss}$ | [ZnL] |
| Control | 55 | 11 | 7 | 200 | 180 | NA |
| 15 mg/L $FeCl_3$ | 58 | 36 | 9 | 160 | 110 | NA |
| 30 mg/L $FeCl_3$ | 79 | 11 | 7 | 180 | 140 | NA |
| 45 mg/L $FeCl_3$ | 96 | 6,8 | 5 | 220 | 120,130 | NA |

| **Treatment** | TSS (mg/L) | pH | $[EDTA]_{total}$ (nM) | $[FeEDTA]_{total}$ (nM) | % FeEDTA |
|---|---|---|---|---|---|
| Control | 10 | 7.29 | 600 | 390 | 65 |
| 15 mg/L $FeCl_3$ | 8.6 | 6.95 | NA | NA | NA |
| 30 mg/L $FeCl_3$ | 17 | 7.38 | 580 | 430 | 74 |
| 45 mg/L $FeCl_3$ | 34 | 7.31 | 600 | 460 | 77 |

NA = analysis not available.

### 3.2.1.6 Summary of Speciation Data for the SR WWTP

For three of the five experiments described in the previous sections, samples were collected after the grit chamber, after primary treatment, and after secondary treatment. These data provide additional information that can be used to assess the effect on EDTA speciation of the pH depression during secondary oxygen-activated sludge treatment.

The data are consistent with previous findings that the percentage of FeEDTA complexes increase during oxygen activated sludge treatment (Table 3-8). On the first date, the percentage of FeEDTA complexes increased from 28% in the influent to 54% in the effluent as the pH dropped during oxygen-activated sludge treatment. On the second date, FeEDTA increased from 53% to 78% between primary and secondary treatment. On the third date, the influent pH was relatively low. As a result, approximately 70% of the EDTA initially consisted of FeEDTA, and little change was observed during treatment.

**Table 3-8. Concentrations of EDTA Complexes at SR WWTP.**

| Sample | pH | Concentration (nM) $[EDTA]_{total}$ | $[NiEDTA^{2-}]$ | $[FeEDTA]_{total}$ | % FeEDTA |
|---|---|---|---|---|---|
| Experiment 2 (5/30/2000) | | | | | |
| 1° Effluent | 7.5 | 740 | 79 | 210 | 28 |
| 2° Effluent | 6.3 | 670 | 82 | 360 | 54 |
| Experiment 4 (7/24/2000) | | | | | |
| 1° Influent | 7.1 | 620 | NA | 370 | 60 |
| 1° Effluent | 7.1 | 600 | NA | 320 | 53 |
| 2° Effluent | 6.7 | 1040 | NA | 810 | 78 |
| Experiment 5 (9/20/2000) | | | | | |
| 1° Influent | 6.9 | 550,650 | NA | 370,500 | 73 |
| 1° Effluent | 7.0 | 530 | NA | 360 | 68 |
| 2° Effluent | NA | 540,560 | NA | 340,360 | 64 |

NA = Not applicable. Results of duplicate treatments are separated with commas.

In conjunction with Experiment 5, samples also were analyzed for $ZnEDTA^{2-}$ and Cu speciation (Figure 3-8). The data suggest that complexation by EDTA affects the removal of Zn at the SR WWTP. As observed previously, concentrations of particle-associated Zn decreased throughout the treatment process while concentrations of dissolved Zn were relatively constant. Comparison of the concentrations of $ZnEDTA^{2-}$ (as measured by the direct HPLC method) with dissolved Zn indicated that 75% to 100% of the dissolved Zn consisted of $ZnEDTA^{2-}$.

In contrast to the results for Zn, EDTA did not play an important role in Cu removal. In the secondary effluent, most of the Cu was associated with particles. The absence of $CuEDTA^{2-}$ complexes (as measured by the HPLC technique) was consistent with the relatively low concentrations of complexed Cu (CuL). Furthermore, the close agreement between the concentration of non-FeEDTA complexes ($[EDTA]_{tot}$-[FeEDTA]) and the concentration of $ZnEDTA^{2-}$ also indicated that $CuEDTA^{2-}$ was not present at concentrations greater than approximately 20 nM.

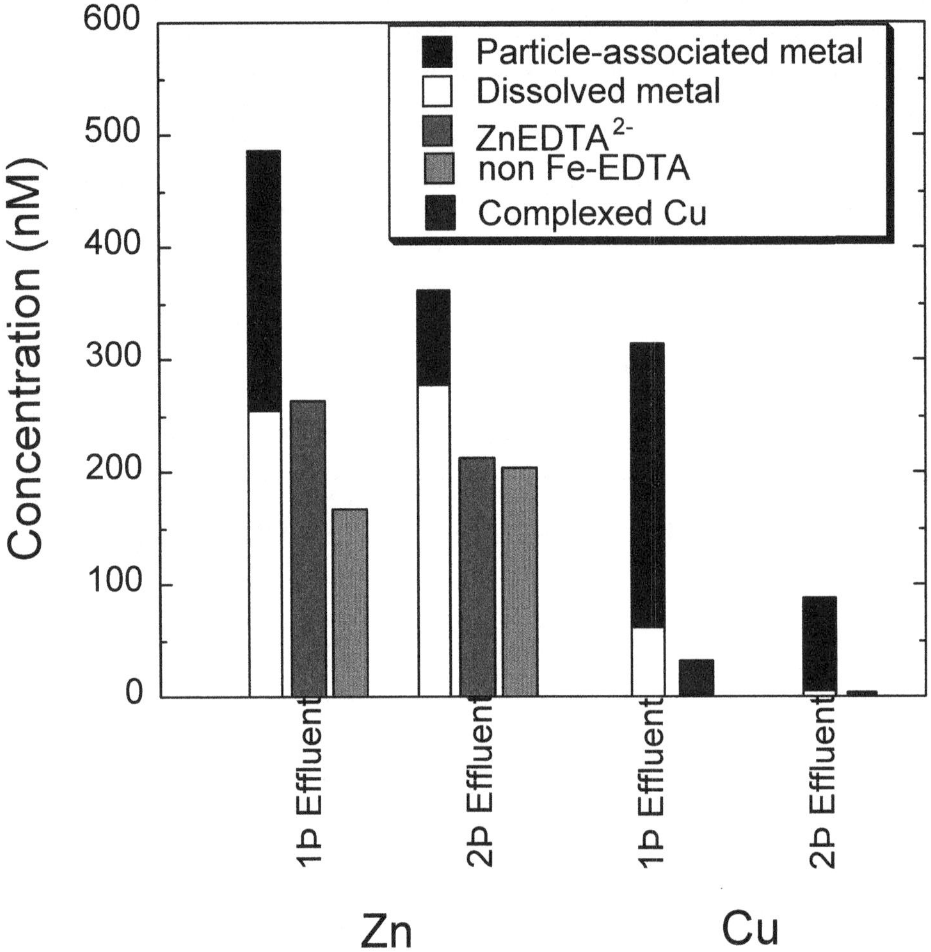


**Figure 3-8. Speciation of Cu and Zn in Primary and Secondary Effluent Samples Collected from SR WWTP on September 20, 2000.**

Overall, results of full-scale measurements at the SR WWTP supported the hypothesis that oxygen-activated sludge treatment results in formation of FeEDTA complexes as $ZnEDTA^{2-}$ and $CuEDTA^{2-}$ dissociate. To illustrate this point, the influent data from the SR WWTP were plotted on the graph of pH versus EDTA speciation (Figure 3-9). To emphasize the new data, we reduced the size of the symbols used to depict the data already included in Figure 2-8. With the exception of the secondary effluent sample collected on May 30, 2000, the EDTA speciation followed the trend observed for the other four WWTPs.

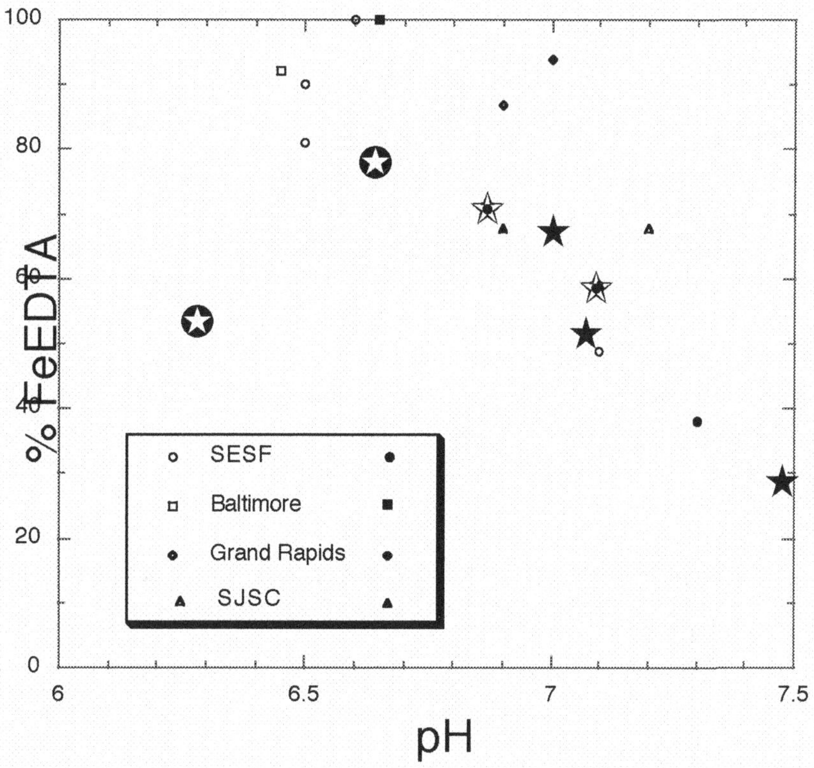


**Figure 3-9. Comparison of Influent EDTA Speciation from the SR WWTP with Data Collected at Other Locations. ✭ indicates primary influent, ★ indicates primary effluent, ✪ indicates secondary effluent.**

### 3.2.2 Millbrae WWTP

As discussed in the previous section, the influent pH at the SR WWTP often was below 7, and as a result, most of the EDTA entering the treatment plant already consisted of FeEDTA complexes. To assess the effects of $FeCl_3$ and acid addition on metal speciation, bench-scale experiments were conducted at the Millbrae WWTP. The influent pH at the Millbrae WWTP is typically greater than 7.5. Four experiments were conducted using conditions similar to those used at the SR WWTP. The first experiment was a preliminary analysis of metal speciation, and the other three experiments replicated the bench-scale experiments described previously. One major difference between the experiments at the SR and Millbrae WWTPs was that the simulated secondary effluent samples were collected in the Millbrae WWTP experiments with a polyethylene pipette as opposed to decantation. This modification avoided the variability in TSS observed in the studies conducted at the SR WWTP.

Overall, results of the three experiments conducted at the Millbrae WWTP indicate that chemical addition during primary treatment could result in a decrease of approximately 20% in effluent Zn concentrations. Addition of $FeCl_3$ with and without concurrent addition of HCl improved the removal of Cu and Zn during simulated primary treatment by reducing suspended solids and dissolved metals. A portion of the $ZnEDTA^{2-}$, which often accounted for a significant fraction of the effluent Zn, dissociated when chemicals were added during simulated primary treatment. Reducing suspended solids may not result in an improvement in metal removal because the solids normally are removed during secondary treatment. However, dissociation of $ZnEDTA^{2-}$ could result in a decrease in Zn concentrations after secondary treatment in a full-scale treatment plant.

### 3.2.2.1 Experiment 1

Prior to conducting bench-scale experiments at the Millbrae WWTP, samples were collected before primary treatment, after primary treatment, and after secondary treatment (Table 3-9, Figure 3-10). Results of these preliminary samples indicated that the Millbrae WWTP has approximately twice the concentration of nonFeEDTA complexes as the SR WWTP. Furthermore, the observation of higher concentrations of nonFeEDTA complexes than the concentration of dissolved Zn suggested that $CuEDTA^{2-}$ may have been present in the plant's influent at significant concentrations. Attempts to measure $CuEDTA^{2-}$ and $ZnEDTA^{2-}$ with the direct HPLC technique failed due to poor post-column conversion and matrix interference. However, the chromatograms were qualitatively consistent with the presence of both $CuEDTA^{2-}$ and $ZnEDTA^{2-}$.

**Table 3-9. Speciation of EDTA and Metal Concentrations in Samples Collected at Millbrae WWTP on October 9, 2000 (Experiment 1).**

| | | Concentration (nM) | | | | |
|---|---|---|---|---|---|---|
| **Treatment** | **pH** | **$[EDTA]_{tot}$** | **[FeEDTA]** | **$[ZnEDTA^{2-}]$** | **$[Cu]_{tot}$** | **$[Cu]_{diss}$** |
| Primary Influent | 7.28 | 620 | 260 | 190 | 650,650 | 220,210 |
| Primary Effluent | 7.34 | 690 | 420 | 370 | 480,470 | 170 |
| Secondary Effluent | 7.30 | 840 | 590 | 80 | 110,110 | 28,31 |

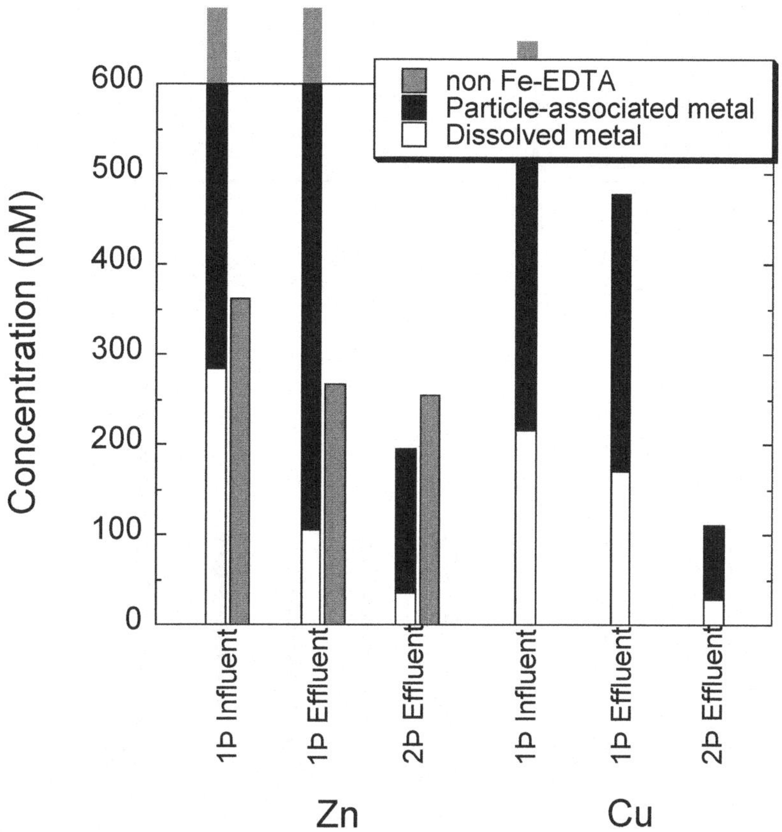


**Figure 3-10. Cu and Zn Measured at Millbrae WWTP on October 9, 2000 (Experiment 1).**

### 3.2.2.2 Experiment 2

In the first bench-scale experiment performed at the Millbrae WWTP, the experimental design employed in study at the SR WWTP was replicated. Results from the first experiment conducted at Millbrae on January 19, 2001 indicated that adding $FeCl_3$ resulted in conversion of a significant percentage of the nonFeEDTA complexes into FeEDTA complexes (Figure 3-11, Table 3-10). Interestingly, the formation of FeEDTA complexes was much more evident after simulated secondary treatment than it was after simulated primary treatment: between primary and secondary treatment the percentage of FeEDTA increased from approximately 40% to 80% in the $FeCl_3$ treatment. For comparison, FeEDTA complexes accounted for less than approximately 40% of the EDTA in the control and pH adjustment treatment.

$FeCl_3$ addition resulted in a small, but measurable, decrease in dissolved concentrations of both metals (Table 3-10, Figure 3-12). After primary treatment, dissolved Cu decreased by 25% and dissolved Zn decreased by 15%, relative to the control. After secondary treatment, dissolved Cu decreased by 15% and dissolved Zn decreased by 5%, relative to controls. The effect of $FeCl_3$ addition on total metal concentrations after primary treatment was not evident, presumably because sample-to-sample variations affected the concentrations of TSS. The variations in TSS were probably an artifact of the experimental procedure and would not have been observed during full-scale treatment (there is no reason why TSS should have increased as a result of $FeCl_3$ addition or moderate pH adjustment).

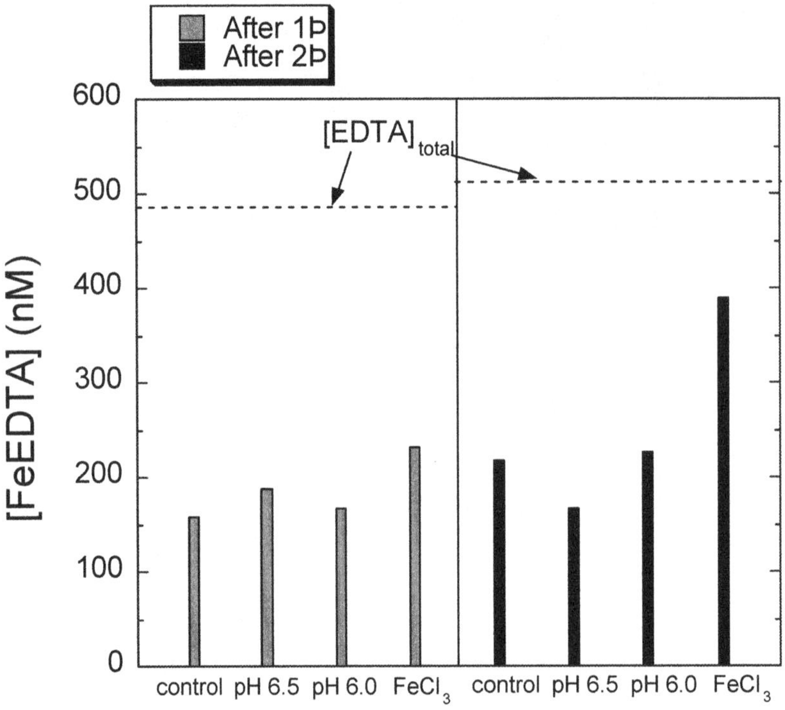


**Figure 3-11. Concentration of FeEDTA Complexes after Simulated Primary Treatment and Simulated Secondary Treatment at Millbrae WWTP on January 19, 2001 (Experiment 2). The average total EDTA concentration in the simulated primary and simulated secondary effluent samples are included for reference.**

**Table 3-10. Results of Samples Collected after Simulated Primary and Secondary Treatment for Chemical Addition to Primary Treatment Experiment Conducted at Millbrae WWTP on January 19, 2001 (Experiment 2).**

**After Primary Treatment**

| | Concentration (nM) | | | | | | |
|---|---|---|---|---|---|---|---|
| **Treatment** | **$Cu_{tot}$** | **$Cu_{diss}$** | **$Zn_{tot}$** | **$Zn_{diss}$** | **EDTA** | **FeEDTA** | **pH[1]** |
| Control | 410,420 | 140,140 | 760, 670 | 300, 330 | 340,500 | 90,230 | 8.28,8.34 |
| pH 6.5 | 430,430 | 130,120 | 790,790 | 440,420 | 530,560 | 200,170 | 6.68,6.64 |
| pH 6.0 | 420 | 130 | 800 | 440 | 490 | 170 | 6.21 |
| 50 mg/L $FeCl_3$ | 440,460 | 100,110 | 700,720 | 270,270 | 500,470 | 230,230 | 7.36,7.2 |

[1] pH measured after completion of treatment.

**After Secondary Treatment**

| | Concentration (nM) | | | | | | |
|---|---|---|---|---|---|---|---|
| **Treatment** | **$Cu_{tot}$** | **$Cu_{diss}$** | **$Zn_{tot}$** | **$Zn_{diss}$** | **EDTA** | **FeEDTA** | **pH[1]** |
| Control | 160,170 | 45,44 | 330,320 | 220,240 | 450,430 | 200,230 | 7.61,7.51 |
| pH 6.5 | 130,160 | 39,31 | 400,390 | 330,290 | 470,490 | 170,170 | 7.43,7.44 |
| pH 6.0 | 160 | 39 | 350 | 270 | 560 | 230 | 7.33 |
| 50 mg/L $FeCl_3$ | 130,130 | 35,41 | 280,300 | 210,230 | 610,590 | 390 | 7.66,7.41 |

[1] pH measured after completion of treatment.

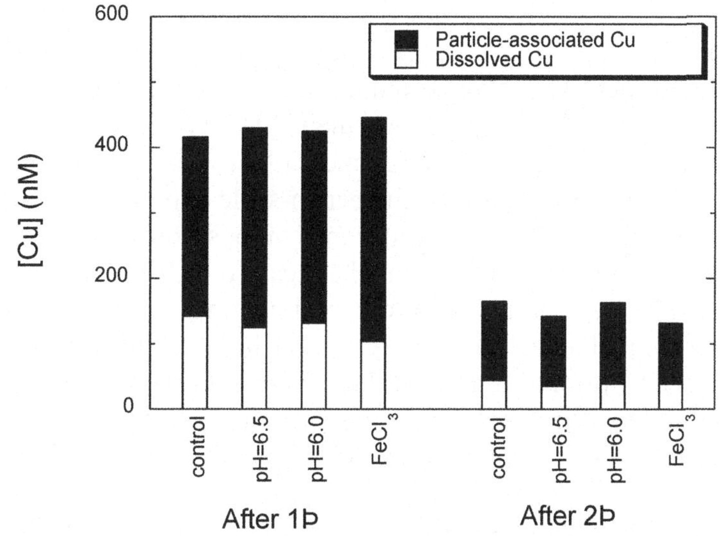


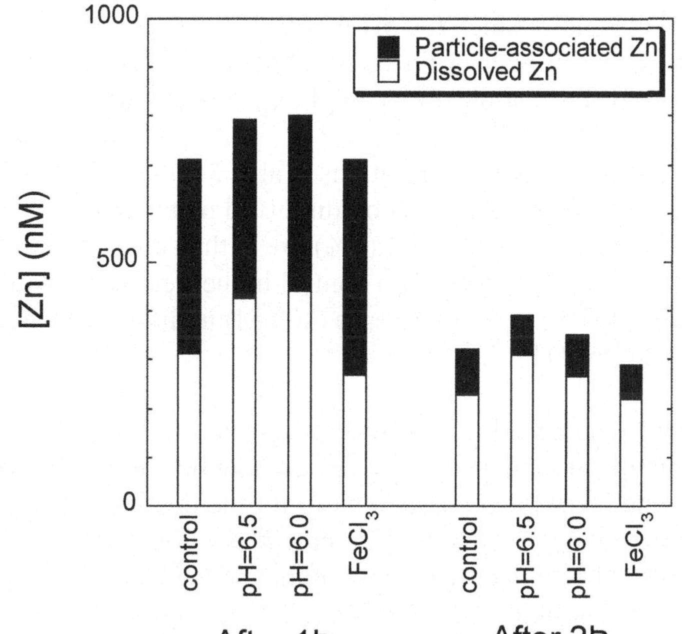


**Figure 3-12. Concentration of Cu and Zn after Simulated Primary Treatment and Simulated Secondary Treatment at Millbrae WWTP on January 19, 2001 (Experiment 2).**

Results of both acid addition treatments indicated that pH adjustment had no effect on Cu. Acidification increased the concentration of dissolved Zn by 35% to 40% relative to control.

The increase in dissolved Zn concentrations, which also has been observed in full-scale treatment systems, was attributable to solubilization of Zn from particles.

**3.2.2.3 Experiment 3**

Results from Experiment 2 suggested that any decrease in metal concentrations following $FeCl_3$ addition would be relatively small (less than 25%). However, as mentioned previously, such a small decrease would be useful to treatment plants operators who are trying to meet metal mass loading requirements. Furthermore, the bench-scale experiments likely underestimate the effect of treatment on effluent quality because the composition of the RAS changes as the mass of metals removed during primary treatment changes. To investigate the effect of treatment on metal removal and EDTA speciation, a third experiment was conducted in which two additional sets of treatments were included. The new treatments involved the addition of either 1,000 nM of $CuEDTA^{2-}$ or $ZnEDTA^{2-}$ prior to treatment. Five different treatments were performed: 1) control; 2) pH adjusted to 6.5; 3) pH adjusted to 6.0; 4) pH adjusted to 5.3; and, 5) addition of 50 mg/L $FeCl_3$. These experiments did not include a simulation of secondary treatment.

Results of EDTA analyses indicated that $FeCl_3$ addition resulted in production of FeEDTA complexes (Figure 3-13, Table 3-11). For the treatment without added EDTA (top of Figure 3-13), FeEDTA increased by approximately 125 nM relative to control. For the treatments with $CuEDTA^{2-}$ and $ZnEDTA^{2-}$, the concentration of FeEDTA complexes increased by 210 nM and 285 nM relative to control. Therefore, between 10% and 20% of the metal-EDTA complexes were transformed into FeEDTA complexes during $FeCl_3$ treatment. With the exception of the treatment in which the pH was adjusted to 5.3, the effect of acid addition on EDTA speciation was very small. Adjustment of pH to 5.3 resulted in production of similar concentrations of FeEDTA complexes as did the treatment with $FeCl_3$.

Measurements of metal concentrations (Table 3-11) indicated that $FeCl_3$ addition had a very small and possibly negligible effect on dissolved metal concentrations. After $FeCl_3$ addition, dissolved Cu concentrations were 35 nM (9%) lower than control in the treatments without added EDTA and 50 nM (5%) lower than control in the treatments with 1,000 nM added $CuEDTA^{2-}$ (Figure 3-14). A similar decrease (40 nM) in dissolved Cu also was observed in the treatment with added $ZnEDTA^{2-}$.

Adding $FeCl_3$ had a negligible effect on concentrations of dissolved Zn (Table 3-11). After treatment, dissolved Zn concentrations were identical to control for the treatment without added $ZnEDTA^{2-}$ and 75 nM lower than control for the treatment with added $ZnEDTA^{2-}$ (Figure 3-14). As was observed in the previous experiment, acidification increased the concentration of dissolved Zn by 10%-20% in the treatment without added $ZnEDTA^{2-}$. Because $FeCl_3$ addition depressed the pH of the solution from 7.95 to 6.81, any positive effect of $FeCl_3$ addition on Zn removal might have been obscured by solubilization of Zn.

Acidification also had a small and possibly negligible effect on dissolved Cu concentrations: when pH was adjusted to 6.5 or 6.0, the concentration of dissolved Cu decreased by less than 20 nM in the treatments without added EDTA and 0 or 50 nM in the treatments with added $CuEDTA^{2-}$. When the pH was adjusted to 5.3, dissolved Cu decreased by 30 and 50 nM in the treatments without and with added $CuEDTA^{2-}$, respectively.

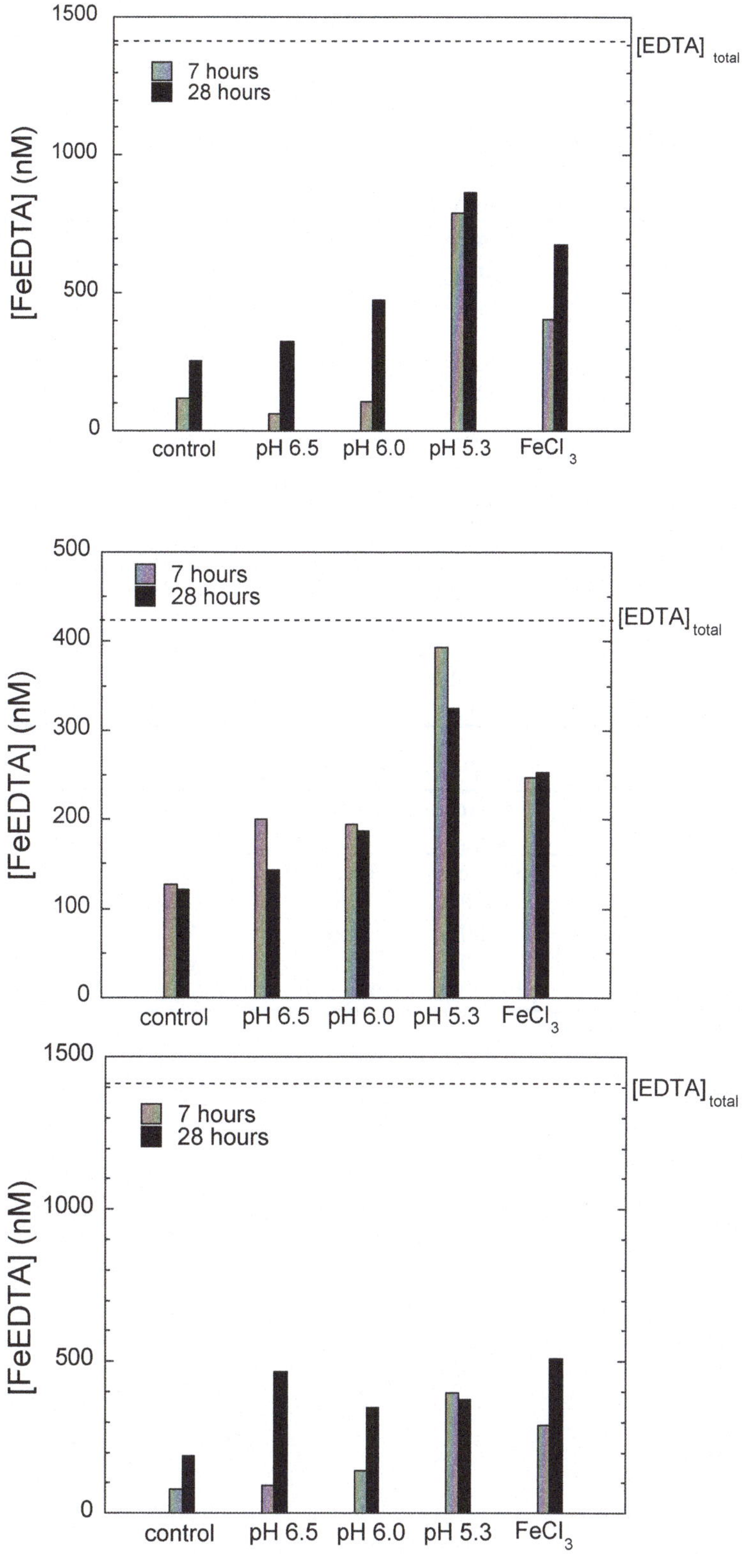


**Figure 3-13. Concentration of FeEDTA Complexes after Simulated Primary Treatment at Millbrae WWTP on February 16, 2001. Top: No added EDTA; Middle: 1,000 nM $CuEDTA^{2-}$ added; 1,000 nM $ZnEDTA^{2-}$ added.**

**Table 3-11. Results of Samples Collected after Simulated Primary Treatment (No Added EDTA) at Millbrae WWTP on February 16, 2001 (Experiment 3).**

**No Added EDTA**

| | Concentration (nM) | | | | | | |
|---|---|---|---|---|---|---|---|
| **Treatment** | **$Cu_{tot}$** | **$Cu_{diss}$** | **$Zn_{tot}$** | **$Zn_{diss}$** | **EDTA** | **FeEDTA** | **pH[1]** |
| Control | 550, 550 | 350, 390 | 810, 820 | 500, 540 | 430,460 | 120,130 | 7.75,7.74 |
| pH 6.5 | 540 | 360 | 830 | 580 | 440 | 200 | 6.63 |
| pH 6.0 | 540 | 350 | 820 | 640 | 380 | 190 | 6.14 |
| pH 5.3 | 530 | 340 | 880 | 720 | 460 | 390 | 5.57 |
| 50 mg/L $FeCl_3$ | 550, 470 | 340, 330 | 820, 760 | 520, 530 | 420,510 | 220,280 | 6.89,6.93 |

**$CuEDTA^{2-}$ Added (1,000 nM)**

| | Concentration (nM) | | | | | | |
|---|---|---|---|---|---|---|---|
| **Treatment** | **$Cu_{tot}$** | **$Cu_{diss}$** | **$Zn_{tot}$** | **$Zn_{diss}$** | **EDTA** | **FeEDTA** | **pH[1]** |
| Control | 1400,1400 | 1200,1100 | 830,840 | 460, 460 | 1200,1100 | 100,60 | 7.76,7.72 |
| pH 6.5 | 1400 | 1200 | 870 | 560 | 1200 | 90 | 6.59 |
| pH 6.0 | 1400 | 1100 | 1000 | 690 | 1200 | 140 | 6.09 |
| pH 5.3 | 1400 | 1100 | 1200 | 830 | 1100 | 340 | 5.60 |
| 50 mg/L $FeCl_3$ | 1300,1300 | 1100,1100 | 800 | 540 | 1200,1300 | 290,290 | 6.84,6.89 |

**$ZnEDTA^{2-}$ Added (1,000 nM)**

| | Concentration (nM) | | | | | | |
|---|---|---|---|---|---|---|---|
| **Treatment** | **$Cu_{tot}$** | **$Cu_{diss}$** | **$Zn_{tot}$** | **$Zn_{diss}$** | **EDTA** | **FeEDTA** | **pH[1]** |
| Control | 520,520 | 370,380 | 1600,1600 | 1000,1100 | 1300,1300 | 160,80 | 7.75,7.77 |
| pH 6.5 | 510 | 380 | 1600 | 1000 | 1300 | 60 | 6.58 |
| pH 6.0 | 520 | 370 | 1600 | 1100 | 1400 | 110 | 6.11 |
| pH 5.3 | 520 | 360 | 1700 | 1100 | 1400 | 790 | 5.53 |
| 50 mg/L $FeCl_3$ | 500,500 | 330,340 | 1400,1400 | 980,970 | 1300,1300 | 430,380 | 6.71,6.80 |

[1] pH measured after completion of treatment.

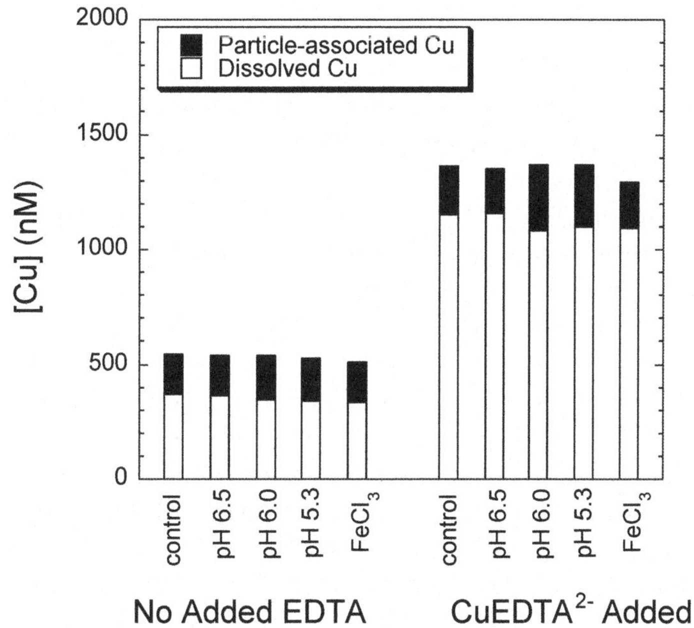


**Figure 3-14. Concentrations of Dissolved and Particle-associated Cu Measured after Simulated Primary Treatment Experiment at Millbrae WWTP on February 16, 2001 (Experiment 3).**

### 3.2.2.4 Experiment 4

The decreases in dissolved Cu and Zn concentrations observed after primary treatment in Experiment 3 were accompanied by a comparable increase in the concentration of FeEDTA complexes. Acidification to pH 6.5 or 6.0 had no effect on Cu or EDTA speciation. Acidification did result in an increase of dissolved Zn by 10% to 20%. To confirm the findings of Experiment 3, the bench-scale experiment at Millbrae was repeated on April 13, 2001. The experiment design was identical to that of Experiment 2 except two additional treatments with 1,000 nM of either $CuEDTA^{2-}$ or $ZnEDTA^{2-}$ were included.

Results of the experiment generally confirmed previous observations. Adding $FeCl_3$ yielded a small but consistent improvement in metal removal after both primary and secondary treatment (Figures 3-15 and 3-16, Table 3-12). Acidification to pH 6.5 had a negligible effect on dissolved Cu and resulted in an increase in dissolved Zn relative to control. Unlike previous experiments, no increase was observed in the concentration of FeEDTA complexes.

Relative to control, adding $FeCl_3$ resulted in a decrease of dissolved Cu after primary treatment of 100 nM (37%) for the sample without added $CuEDTA^{2-}$ (Figure 3-15). After secondary treatment, the concentration of dissolved Cu decreased by 47 nM (36%) relative to control, while the treatment with added $CuEDTA^{2-}$ increased by 20 nM (3%) relative to control, suggesting that no added $CuEDTA^{2-}$ was removed by $FeCl_3$ addition. Therefore, the decrease in dissolved Cu measured in the experiment without added Cu could have been attributable to removal of some other species. Overall, $FeCl_3$ addition decreased the total concentration of Cu by 120 nM after primary treatment and 30 nM after secondary treatment relative to control in the experiment without added $CuEDTA^{2-}$

Acidification had a negligible effect on dissolved Cu relative to control: in the experiment without added EDTA there was no difference between dissolved Cu concentrations for the control and the acidification treatment. With added $CuEDTA^{2-}$, dissolved Cu was 100 nM higher than control following acidification. After secondary treatment, the concentration of dissolved Cu was 30 nM lower without added $CuEDTA^{2-}$ and 10 nM lower than control with added $CuEDTA^{2-}$. Overall, total Cu concentrations were identical after primary treatment and 10 nM higher after secondary treatment without added EDTA. Total Cu concentrations were identical to those measured in the control after primary treatment and 70 nM lower with added $CuEDTA^{2-}$, indicating that adjusting the pH to 6.5 had no effect on Cu removal.

$FeCl_3$ addition reduced the concentration of dissolved Zn after primary and secondary treatment in the experiments with and without added $ZnEDTA^{2-}$ (Figure 3-16) After primary treatment, the concentration of dissolved Zn was 110 nM (31%) lower without added EDTA and 150 nM (15%) lower with added $ZnEDTA^{2-}$. After secondary treatment, the concentration of dissolved Zn was 50 nM (19%) lower than that of the control without added EDTA and 100 nM (13%) lower with added $ZnEDTA^{2-}$. Overall, total Zn concentrations were 230 nM (34%) lower than control after primary treatment and 70 nM (22%) lower than control after secondary treatment without added EDTA. Total Zn concentrations were 330 nM (34%) lower than control after primary treatment and 120 nM (14%) lower than control after secondary treatment with added $ZnEDTA^{2-}$. Adjusting pH resulted in higher concentrations of Zn after primary treatment and comparable or slightly higher Zn concentrations after secondary treatment. In the samples that were not amended with $ZnEDTA^{2-}$, adjusting pH to 6.5 increased dissolved Zn concentrations by 56% after primary treatment and by 24% after secondary treatment (Figure 3-16). The effect of pH adjustment on dissolved Zn was smaller in the experiment conducted after amendment with $ZnEDTA^{2-}$ (increases of 21% after primary and 7% after secondary treatment) because the source of the dissolved Zn was particulate matter, which is unaffected by $ZnEDTA^{2-}$ addition. When the absolute increase in dissolved Zn in both treatments relative to control was compared, similar increases were observed after primary treatment (195 nM without added $ZnEDTA^{2-}$ and 200 nM with added $ZnEDTA^{2-}$). Therefore, pH adjustment results in solubilization of particle-associated Zn but has no effect on $ZnEDTA^{2-}$.

Measurement of EDTA concentrations and speciation indicated that 130 nM of $CuEDTA^{2-}$ and 100 nM of $ZnEDTA^{2-}$ dissociated when EDTA was added to the wastewater. No dissociation of EDTA complexes was observed without added EDTA complexes. However, interference from organic matter decreased the precision of the analysis. In the treatments conducted without adding $ZnEDTA^{2-}$, the total concentration of EDTA was between 500 and 600 nM, of which approximately 50% to 60% consisted of FeEDTA complexes (Table 3-12). In the experiment with 1,000 nM of added $ZnEDTA^{2-}$, the total concentration of EDTA increased to approximately 1,500 nM after primary treatment, which indicates that little, if any, of the added EDTA was removed during primary treatment. The slightly lower concentrations of EDTA measured after secondary treatment are attributable to dilution with return activated sludge.

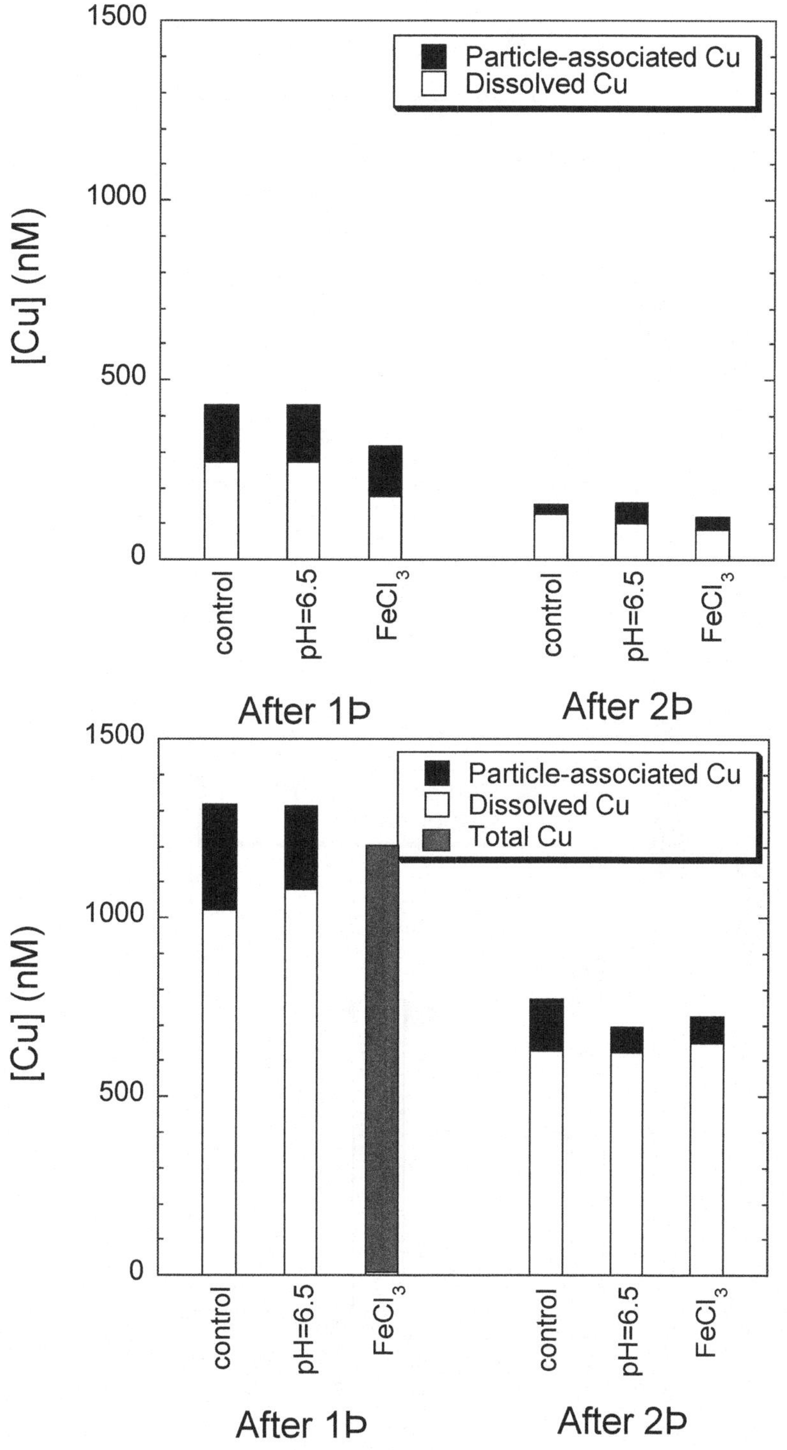


**Figure 3-15. Concentrations of Dissolved and Particle-Associated Cu Measured after Simulated Primary and Simulated Secondary Treatment Experiment at Millbrae WWTP on April 13, 2001 (Experiment 4). Top: No added EDTA; Bottom: 1,000 nM $CuEDTA^{2-}$ added prior to treatment.**

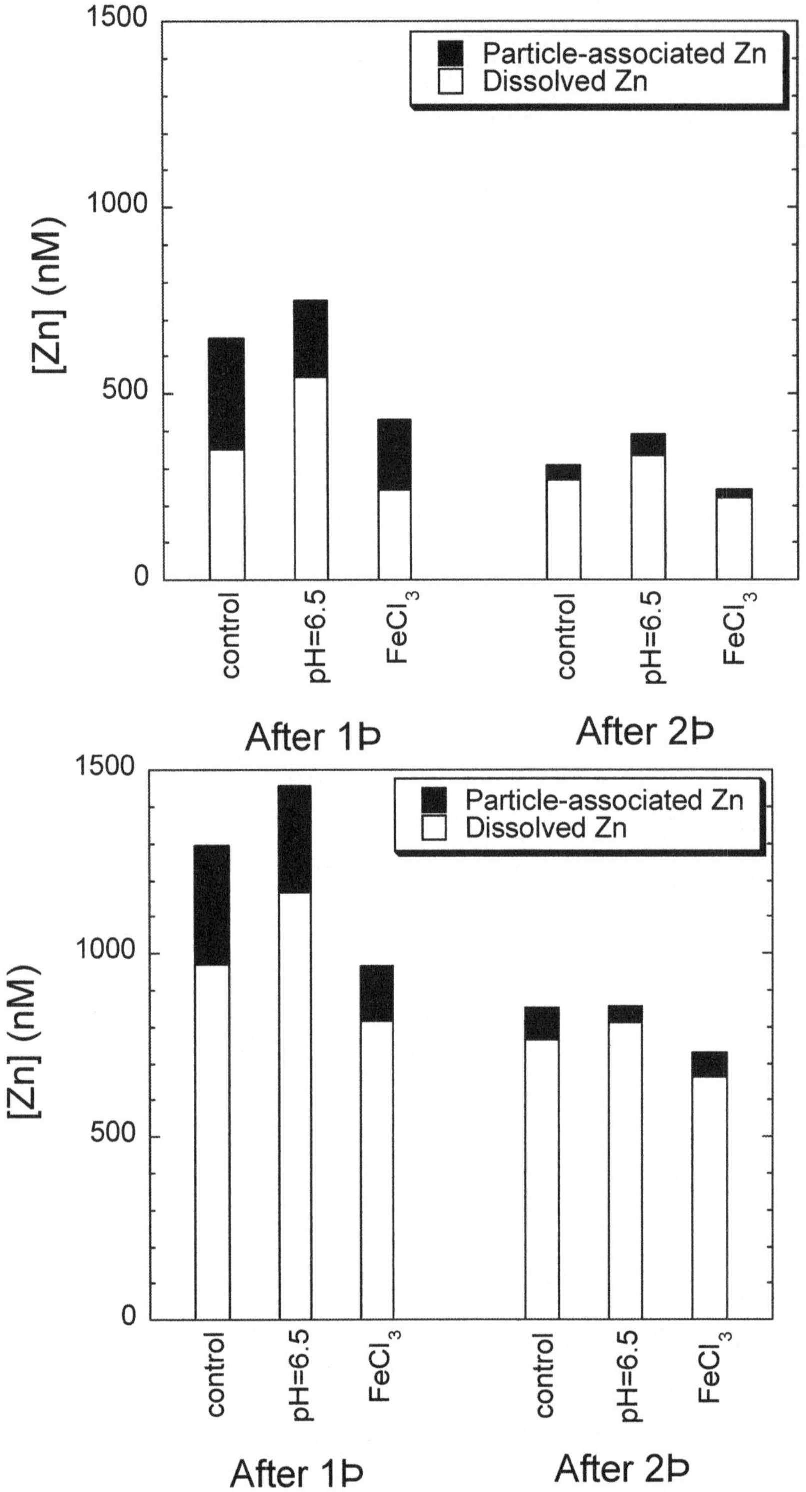


**Figure 3-16. Concentrations of Dissolved and Particle-Associated Zn Measured after Simulated Primary and Simulated Secondary Treatment Experiment at Millbrae WWTP on April 13, 2001 (Experiment 4). Top: No added EDTA; Bottom: 1,000 nM $ZnEDTA^{2-}$ added prior to treatment.**

**Table 3-12. Results of Samples Collected after Simulated Primary and Secondary Treatment (no added EDTA) at Millbrae WWTP on April 13, 2001 (Experiment 4).**

**After Primary Treatment**

| | Concentration (nM) | | | | | | | | Concentration (mg/L) | | |
|---|---|---|---|---|---|---|---|---|---|---|---|
| **Treatment** | $Cu_{tot}$ | $Cu_{diss}$ | $Cu_{labile}$ | $Zn_{tot}$ | $Zn_{diss}$ | $Zn_{labile}$ | EDTA | FeEDTA | DOC | TSS | pH[1] |
| Control | 430 | 270 | 18 | 650 | 350 | 63 | 580, 530 | 350, 380 | 21 | 68 | 7.88 |
| pH 6.5 | 430 | 270 | 13 | 750 | 550 | 200 | 590 | 230 | 25 | 71 | 6.60 |
| 50 mg/L $FeCl_3$ | 310 | 170 | 12 | 430 | 240 | 48 | 520 | 330 | 22,22 | 59 | 7.05 |

**After Secondary Treatment**

| | Concentration (nM) | | | | | | | | Concentration (mg/L) | | |
|---|---|---|---|---|---|---|---|---|---|---|---|
| **Treatment** | $Cu_{tot}$ | $Cu_{diss}$ | $Cu_{labile}$ | $Zn_{tot}$ | $Zn_{diss}$ | $Zn_{labile}$ | EDTA | FeEDTA | DOC | TSS | pH |
| Control | 150 | 130 | 7.1 | 310 | 270 | 59 | 570, 610 | 340 | 18 | 14 | 7.88 |
| pH 6.5 | 160 | 100 | 10 | 390 | 330 | 110 | 570 | 370 | 19 | 14 | 7.67 |
| 50 mg/L $FeCl_3$ | 120 | 83 | 6.8 | 240 | 220 | 80 | 590 | 400 | 12 | 12 | 7.81 |

Notes:
NA = analysis not available.
[1] pH measured after completion of treatment.

**Results of Samples Collected after Simulated Primary and Secondary Treatment (1,000 nM $CuEDTA^{2-}$ added) at Millbrae WWTP April 13, 2001.**

**After Primary Treatment**

| | Concentration (nM) | | | | | | | | Concentration (mg/L) | | |
|---|---|---|---|---|---|---|---|---|---|---|---|
| **Treatment** | $Cu_{tot}$ | $Cu_{diss}$ | $Cu_{labile}$ | $Zn_{tot}$ | $Zn_{diss}$ | $Zn_{labile}$ | EDTA | FeEDTA | DOC | TSS | pH[1] |
| Control | 1300 | 1000 | 120 | 640 | 360 | 24 | 1300 | 150 | NA | 77 | 7.88 |
| pH 6.5 | 1300 | 1100 | 140 | 1100 | 920 | 230 | 1500 | 22 | 27 | 89 | 6.54 |
| 50 mg/L $FeCl_3$ | 1200 | NA | 110 | 400 | 360 | 25 | 1400 | 280 | NA | 68 | 7.06 |

**After Secondary Treatment**

| | Concentration (nM) | | | | | | | | Concentration (mg/L) | | |
|---|---|---|---|---|---|---|---|---|---|---|---|
| **Treatment** | $Cu_{tot}$ | $Cu_{diss}$ | $Cu_{labile}$ | $Zn_{tot}$ | $Zn_{diss}$ | $Zn_{labile}$ | EDTA | FeEDTA | DOC | TSS | pH |
| Control | 770 | 630 | 62 | 320 | 310 | 43 | 1400 | 230 | NA | 12 | 7.90 |
| pH 6.5 | 700 | 620 | 72 | 550 | 550 | 120 | 1400 | 340 | NA | 9 | 7.71 |
| 50 mg/L $FeCl_3$ | 730 | 650 | 59 | 270 | 250 | 60 | 1500 | 340 | NA | 9 | 7.85 |

Notes:
NA = analysis not available.
[1] pH measured after completion of treatment.

**Table 3-12 (continued). Results of Samples Collected after Simulated Primary and Secondary Treatment (1,000 nM $ZnEDTA^{2-}$ added) at Millbrae WWTP on April 13, 2001.**

**After Primary Treatment**

| Treatment | Concentration (nM) | | | | | | | | Concentration (mg/L) | | |
|---|---|---|---|---|---|---|---|---|---|---|---|
| | $Cu_{tot}$ | $Cu_{diss}$ | $Cu_{labile}$ | $Zn_{tot}$ | $Zn_{diss}$ | $Zn_{labile}$ | EDTA | FeEDTA | DOC | TSS | pH[1] |
| Control | 430 | 280 | 23 | 1300 | 970 | 74 | 1400 | 140 | NA | 75 | 7.90 |
| pH 6.5 | 440 | 240 | 14 | 1500 | 1200 | 1100 | 1300 | <50 | NA | 81 | 6.57 |
| 50 mg/L $FeCl_3$ | 310 | 170 | 12 | 970 | 820 | 120 | 1400 | 240 | 22 | 64 | 7.06 |

**After Secondary Treatment**

| Treatment | Concentration (nM) | | | | | | | | Concentration (mg/L) | | |
|---|---|---|---|---|---|---|---|---|---|---|---|
| | $Cu_{tot}$ | $Cu_{diss}$ | $Cu_{labile}$ | $Zn_{tot}$ | $Zn_{diss}$ | $Zn_{labile}$ | EDTA | FeEDTA | DOC | TSS | pH |
| Control | 150 | 120 | 6.3 | 850 | 760 | 85 | 1300 | 260 | NA | 12 | 7.85 |
| pH 6.5 | 170 | 110 | 6.9 | 850 | 810 | 95 | 1300 | 88 | NA | 12 | 7.70 |
| 50 mg/L $FeCl_3$ | 110 | 90 | 9.6 | 730 | 660 | 94 | 1300 | 230 | 13 | 12 | 7.80 |

Notes:
NA = analysis not available.
[1] pH measured after completion of treatment.

### 3.2.2.5 Summary of Speciation Data for the Millbrae WWTP

The effects of $FeCl_3$ addition on dissolved metal concentrations and EDTA speciation for the three experiments described in this progress report are summarized in Table 3-13. The first column for each variable indicates the difference between the $FeCl_3$ treatment and the control measured after primary treatment. These results indicate that $FeCl_3$ treatment resulted in a decrease in the concentration of dissolved Cu of 35-110 nM. The absolute magnitude of this effect did not appear to depend on whether or not $CuEDTA^{2-}$ had been added. Results for Zn indicated that $FeCl_3$ addition decreased the concentration of dissolved Zn in four of the seven experiments. Dissolved Zn concentrations were nearly identical to control for two of the treatments (Experiments 3A and 4B) and were higher in one experiment (Experiment 3B). Two of the experiments in which dissolved Zn concentrations increased (Experiments 3B and 4B) should not be considered as indicative of the fate of Zn in real treatment systems because they involved the addition of $CuEDTA^{2-}$, which could have been converted into $ZnEDTA^{2-}$ during treatment. It is possible that the pH decrease that occurred after $FeCl_3$ addition counteracted the effect of dissociation of $ZnEDTA^{2-}$.

**Table 3-13. Summary of Effects of Ferric Chloride Addition on Concentrations of Dissolved Metals and EDTA Speciation Compared to Control.**

| | Δ Cu[1] | | Δ Zn | | Δ FeEDTA | |
|---|---|---|---|---|---|---|
| Experiment | Concn (nM) | % | Concn (nM) | % | Concn (nM) | % EDTA[2] |
| 2 | -35 | -25 | -45 | -14 | +70 | 47 |
| 3A | -35 | -9.4 | +5 | +1 | +125 | 54 |
| 3B ($CuEDTA^{2-}$) | -50 | -4.8 | +80 | +17 | +210 | 23 |
| 3C ($ZnEDTA^{2-}$) | -40 | -11 | -75 | -7.1 | +285 | 27 |
| 4A | -100 | -37 | -110 | -31 | -40 | 63 |
| 4B ($CuEDTA^{2-}$) | NA | NA | 0 | 0 | +130 | 20 |
| 4C ($ZnEDTA^{2-}$) | -110 | -39 | -150 | -15 | +100 | 17 |

[1] Indicates the difference between dissolved metal concentration after primary treatment compared to control:
$\Delta Me = [Me]_{diss,FeCl_3} - [Me]_{diss,control}$

[2] $$\%EDTA = \frac{[FeEDTA]_{FeCl_3}}{[EDTA]_{FeCl_3}}$$

In six of the seven experiments, the concentration of FeEDTA complexes increased after $FeCl_3$ addition. In the experiment in which the concentration of FeEDTA decreased (Experiment 4A) the percent of the EDTA that consisted of FeEDTA complexes was higher than that measured in the other experiments in which EDTA had not been added. Also noteworthy is that the concentration of FeEDTA complexes measured in the acidified treatment conducted during Experiment 4B was lower than that measured in the control. Because acidification did not alter the speciation of EDTA in any of the other samples, the data for these samples may be erroneous.

In summary, $FeCl_3$ addition during primary treatment resulted in lower concentrations of dissolved Cu and, in most experiments, lower concentrations of Zn relative to control. Some of the decrease was probably attributable to conversion of $CuEDTA^{2-}$ and $ZnEDTA^{2-}$ into FeEDTA complexes, followed by removal of the Cu and Zn released from the EDTA. However, some of the removal also could have been due to removal of other Cu and Zn species.

CHAPTER 4.0

# FULL-SCALE STUDY TO ASSESS THE EFFECT OF CHEMICAL ADDITION DURING PRIMARY TREATMENT ON METAL REMOVAL

## 4.1 Background

Results from the bench-scale studies described in Chapter 3.0 suggested that addition of $FeCl_3$ would result in a decrease in the concentrations of Cu and Ni in wastewater effluent by approximately 20%. The improved metal removal appeared to be related to the conversion of pollutant metal-EDTA complexes into FeEDTA complexes and improved removal of particles and uncomplexed metals. Despite these promising results, the efficacy of the process is uncertain because the conditions used in the bench-scale experiments were only a crude surrogate for an actual treatment plant. Differences were expected between the bench-scale experiment and the full-scale treatment process because the bench-scale reactors operated in a batch mode rather than a continuous mode, and the RAS composition presumably would change if $FeCl_3$ were not added to the wastewater. Furthermore, the dose of $FeCl_3$ used in the bench-scale experiments (15-50 mg/L) was significantly higher than that used at the full-scale treatment plant (10 mg/L).

To verify that the addition $FeCl_3$ during primary treatment lowers the concentrations of Cu and Zn metals in wastewater effluent, an experiment was performed at the Millbrae WWTP, which was chosen because it currently adds $FeCl_3$ during primary treatment. Therefore, the effect of $FeCl_3$ could be studied by interrupting and resuming its use.

## 4.2 WWTP Description and Experimental Design

The study was conducted between July and October 2001 at the Millbrae (Calif.) municipal WWTP, an air-activated sludge treatment plant treating an average dry weather flow of 0.07 $m^3/s$ (1.7 MGD). At the Millbrae WWTP, treatment consists of influent grinding, primary settling without initial grit removal, air-activated sludge secondary treatment, secondary clarification, and effluent chlorination. Effluent is dechlorinated offsite before being discharged to San Francisco Bay. $FeCl_3$ is normally added to plant influent at a concentration of 10 mg $FeCl_3$/L just prior to primary settling. As part of this study, the plant's normal $FeCl_3$ addition was interrupted for a period of 35 days to allow comparison of plant performance with and without $FeCl_3$ addition. Eleven grab samples were collected during an initial 33-day period of normal $FeCl_3$ addition, 11 grab samples were collected during a 35-day period of suspended $FeCl_3$ addition, and eight samples were collected during a 19-day period immediately after $FeCl_3$ addition was resumed. Samples were collected from upstream of the $FeCl_3$ addition point, from the effluent of primary treatment, and from the secondary clarifiers.

Grab samples were collected starting with the secondary effluent (secondary effluent typically was collected at 10:30 a.m., primary effluent at 11:15 a.m., and primary influent at noon). Because the same procedure was followed during the periods with and without $FeCl_3$ addition, diurnal variations should not introduce bias into the evaluation.

Grab samples for pH measurement were filtered through an in-line Gelman Aqua Prep 600 0.45 μM cartridge filter into 120-mL sample bottles that were filled to approximately 60 mL with 60 mL headspace in order to avoid pH depression due to oversaturation of the headspace with carbon dioxide. Measurements of pH were made within 1.5 hours of sample collection and were repeated approximately 6 hours after sample collection (after return to the laboratory). TSS of grab samples and of 24-hour composite samples of primary influent, primary effluent, and secondary effluent were measured according to standard methods (APHA 1998). Grab samples for EDTA and metals analysis were collected with a peristaltic pump fitted with acid-cleaned Bev-a-Line (Cole-Parmer) polyethylene-lined tubing using trace metal clean techniques. Samples were stored in acid-cleaned 60-mL bottles. Samples for dissolved metals or EDTA analysis were filtered during sampling with an in-line Gelman Aqua Prep 600 0.45-μm cartridge filter. Samples used for EDTA measurements were filled half full to avoid pH depression. Samples for metals analysis were filled completely to minimize the potential for contamination. Samples for EDTA measurements were kept in a cooler in the field and were refrigerated at 4°C within 4.5 hours after sample collection. Samples were analyzed for dissolved and total metals, total EDTA, and EDTA speciation using the methods described in Chapter 3.0. Other water quality parameters, including pH and TSS, were measured as described previously.

## 4.3 Results

### 4.3.1 Total Suspended Solids

No relationship was observed between $FeCl_3$ addition and TSS in 24-hour composite samples of primary or secondary effluent (Figure 4-1). The absence of a relationship between TSS and $FeCl_3$ addition suggested that the $FeCl_3$ addition had little effect on the efficiency of the primary treatment process. TSS in primary effluent samples ranged from 75–150 mg $L^{-1}$. In secondary effluent samples, TSS typically ranged from 5–10 mg/L with the exception of a period after a plant upset that occurred on Day 51 of the experiment. The upset resulted in an increase of secondary effluent TSS to approximately 25 mg/L, which returned to normal approximately 18 days later. Plant operators attributed the upset to a festival held on Labor Day weekend in the City of Millbrae.

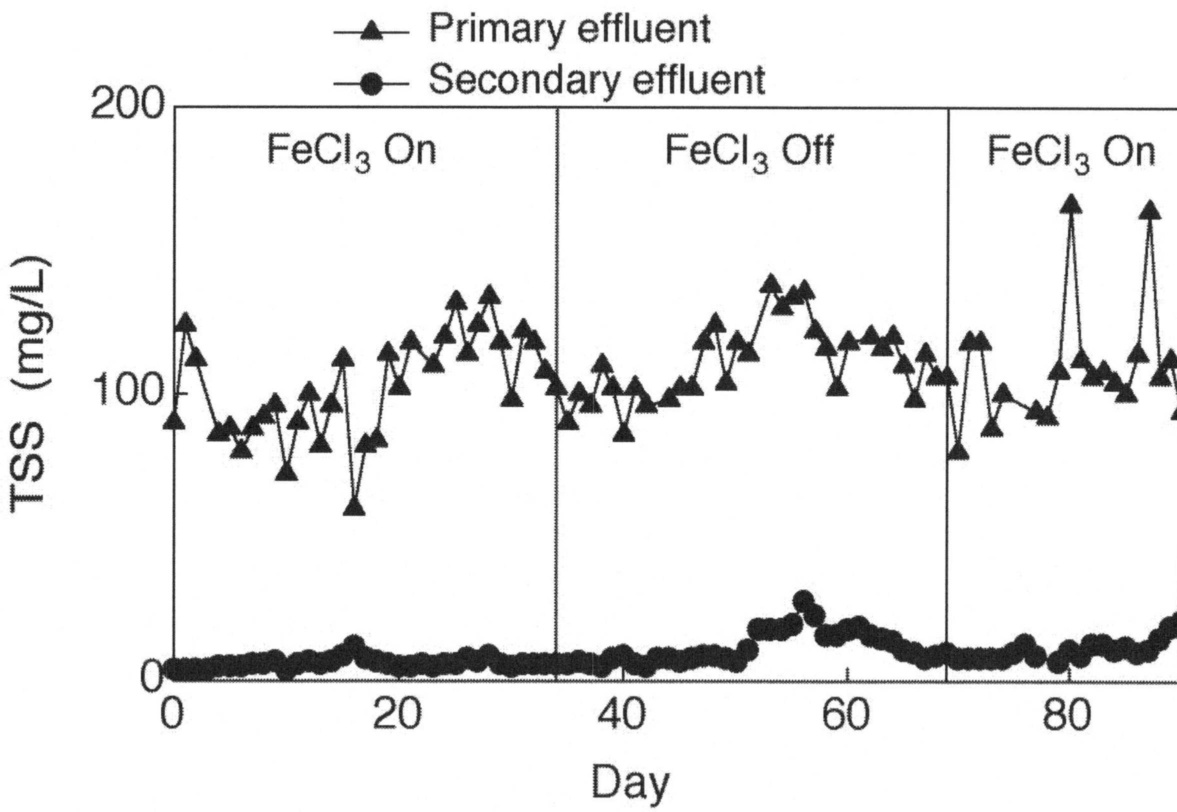


**Figure 4-1. Total Suspended Solids of 24-hour Composite Samples of Primary and Secondary Effluent during the Study Period. Vertical bars indicate days on which ferric chloride addition was started or stopped.**

### 4.3.2 Iron Concentrations

As expected, total Fe concentrations in primary effluent samples decreased during the period when $FeCl_3$ addition was discontinued (Figure 4-2, top). $FeCl_3$ addition did not have a significant effect on primary effluent dissolved Fe concentrations because primary effluent was oversaturated with Fe(III) prior to $FeCl_3$ addition (the solubility of Fe(III) is much less than 1 μM under at pH values above 6). Similarly, the concentration of total and dissolved Fe in secondary effluent remained relatively constant throughout the study period (Figure 4-2, bottom). The two unusually high total Fe concentrations detected in secondary effluent samples collected on Days 20 and 34 likely were caused by the presence of iron-rich particles in the two affected samples.

### 4.3.3 EDTA Speciation and Concentration

As discussed previously, EDTA in municipal wastewater consists mainly of FeEDTA complexes and EDTA complexes with Zn, Cu, and Ni. Thus, the total concentration of EDTA that is not accounted for by FeEDTA, referred to here as nonFeEDTA, should be a measure of the sum of all pollutant-metal EDTA complexes. At the Millbrae WWTP, nonFeEDTA in secondary effluent often was largely composed of $ZnEDTA^{2-}$. The concentration of $ZnEDTA^{2-}$, as measured by the ion exchange chromatography method, accounted for 86 ± 45% (mean ± s.d.) of nonFeEDTA and 81 ± 31% (mean ± s.d.) of dissolved Zn in secondary effluent samples (Figure 4-3).

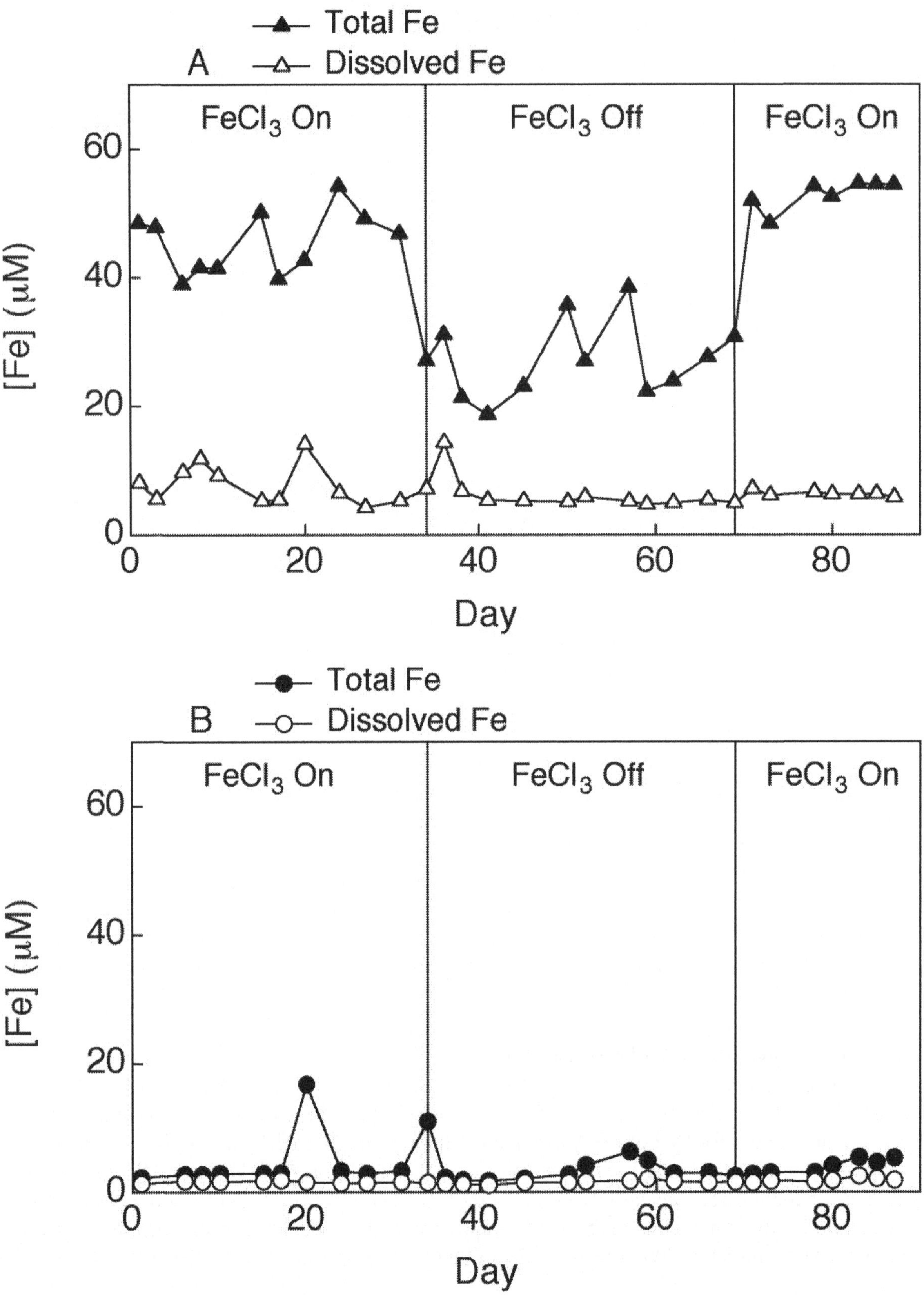


**Figure 4-2. Total and Dissolved Fe in Primary Effluent (top) and Secondary Effluent (Bottom) Samples Collected at Millbrae WWTP from July-October 2001.**

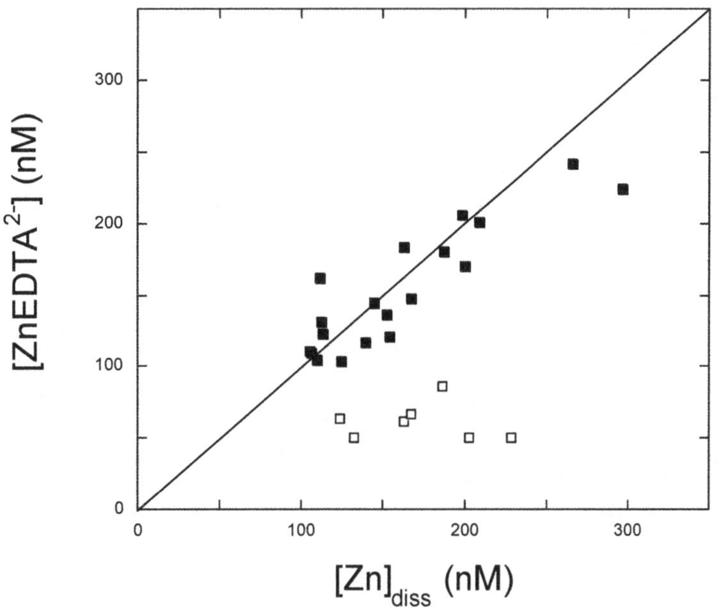


**Figure 4-3. Comparison of $ZnEDTA^{2-}$ and Dissolved Zn in Secondary Effluent Samples Collected at Millbrae WWTP. The line indicates a 1:1 relationship. The hollow squares indicate $ZnEDTA^{2-}$ measurements below the method detection limit.**

The total concentration of EDTA remained relatively constant during the treatment process (Figure 4-4). Although a few samples contained concentrations that were more than twice as high as the median values, differences among the concentrations of EDTA in the raw sewage, primary effluent, and secondary effluent were not statistically significant. These results were consistent with previous studies (Alder et al., 1990; Kari and Giger, 1996) that showed little or no removal of EDTA in municipal wastewater treatment plants.

Because the total concentration of EDTA was not affected by treatment, and EDTA concentrations measured at different locations within the plant varied by approximately 30% on any given day, it was easier to detect trends in EDTA speciation by examining the percentage of EDTA that was associated with Fe ($100 * \frac{[FeEDTA]}{[EDTA]_{total}}$). In this situation, a decrease in the concentration of nonFeEDTA, accompanied by an increase in the concentration of FeEDTA, indicated conversion of pollutant-metal EDTA into FeEDTA complexes. Plotting the data from the entire study in this manner indicated that pollutant metal-EDTA complexes were converted into FeEDTA complexes during primary treatment and that little change in EDTA speciation occurred during secondary treatment (Figure 4-5). During the entire study period, the mean percentage of FeEDTA in all primary influent samples was 35±3%, increasing to a mean of 61±3% in primary effluent samples. On average, primary effluent samples had 160 nM higher concentrations of FeEDTA complexes and 200 nM lower concentrations of nonFeEDTA than primary influent samples. No difference was detected between the mean percentage FeEDTA in primary effluent samples 61± 3% and secondary effluent samples (65± 2%).

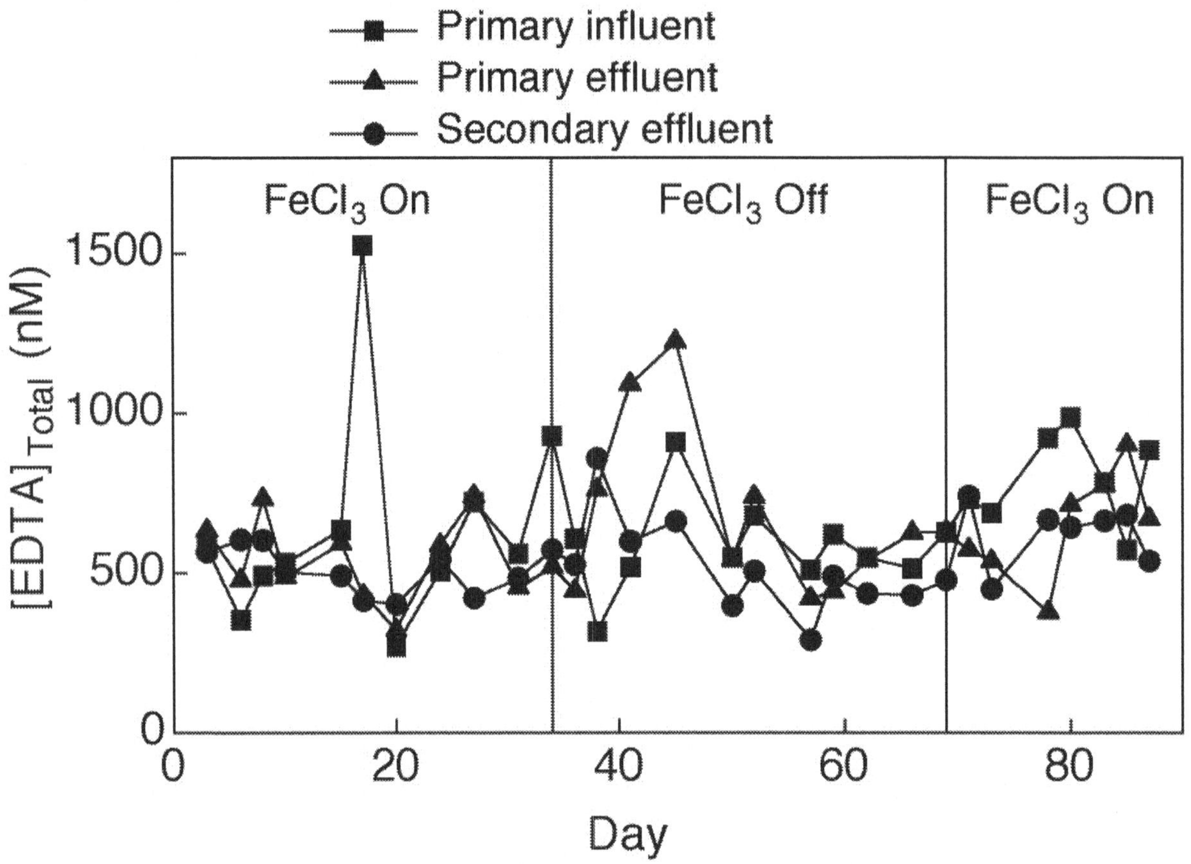


**Figure 4-4. Total Concentrations of EDTA Measured at Millbrae WWTP.**

The mean value of the percentage FeEDTA and the standard deviation on the mean in secondary effluent was 68±2 (median=70%) during the period with $FeCl_3$ addition and 60±5% (median=59%) during the period without $FeCl_3$ addition. In samples of secondary effluent, the mean concentration of FeEDTA was lower during the period without $FeCl_3$ addition (305 ± 28 nM) as compared to the period with $FeCl_3$ addition (371 ± 24 nM). The mean concentration of nonFeEDTA was higher during the period without $FeCl_3$ addition (219 ± 38 nM) as compared to the period with $FeCl_3$ addition (167 ± 13 nM).

To determine if the changes in EDTA speciation observed during primary treatment were related to the presence of particles, the percentage of FeEDTA was measured in samples of primary influent stored at room temperature (Figure 4-6). During storage, the percentage of FeEDTA increased in samples of unfiltered primary influent, while no change in the percentage FeEDTA was observed in samples that had been filtered at the time of collection. These results suggest that particles play an important role in the conversion of pollutant metal-EDTA complexes into FeEDTA complexes.

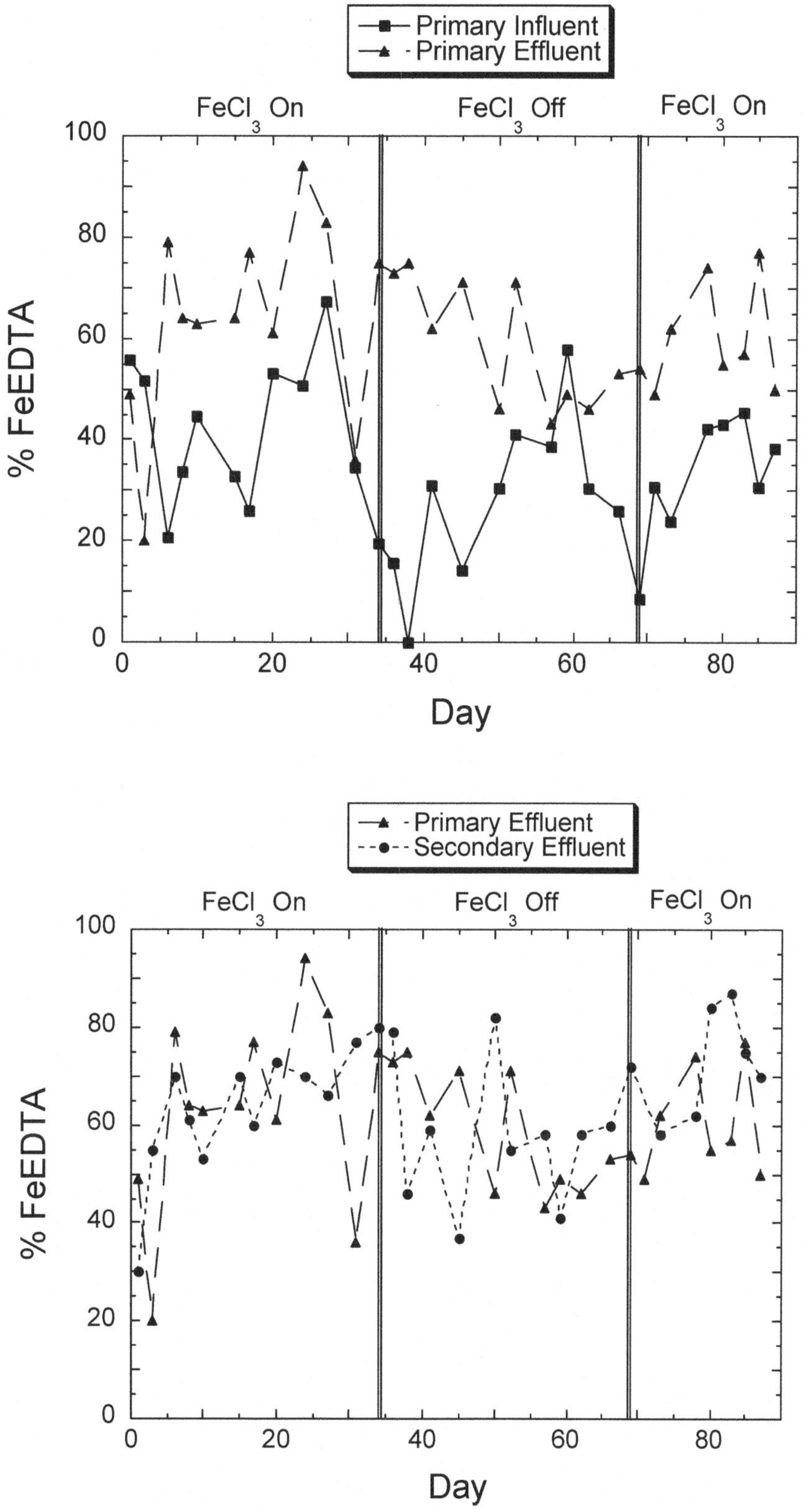


**Figure 4-5. Percentage of EDTA Complexed by Fe Measured at Millbrae WWTP. An increase in the % FeEDTA was O\observed during primary treatment (top) but not during secondary treatment (bottom).**

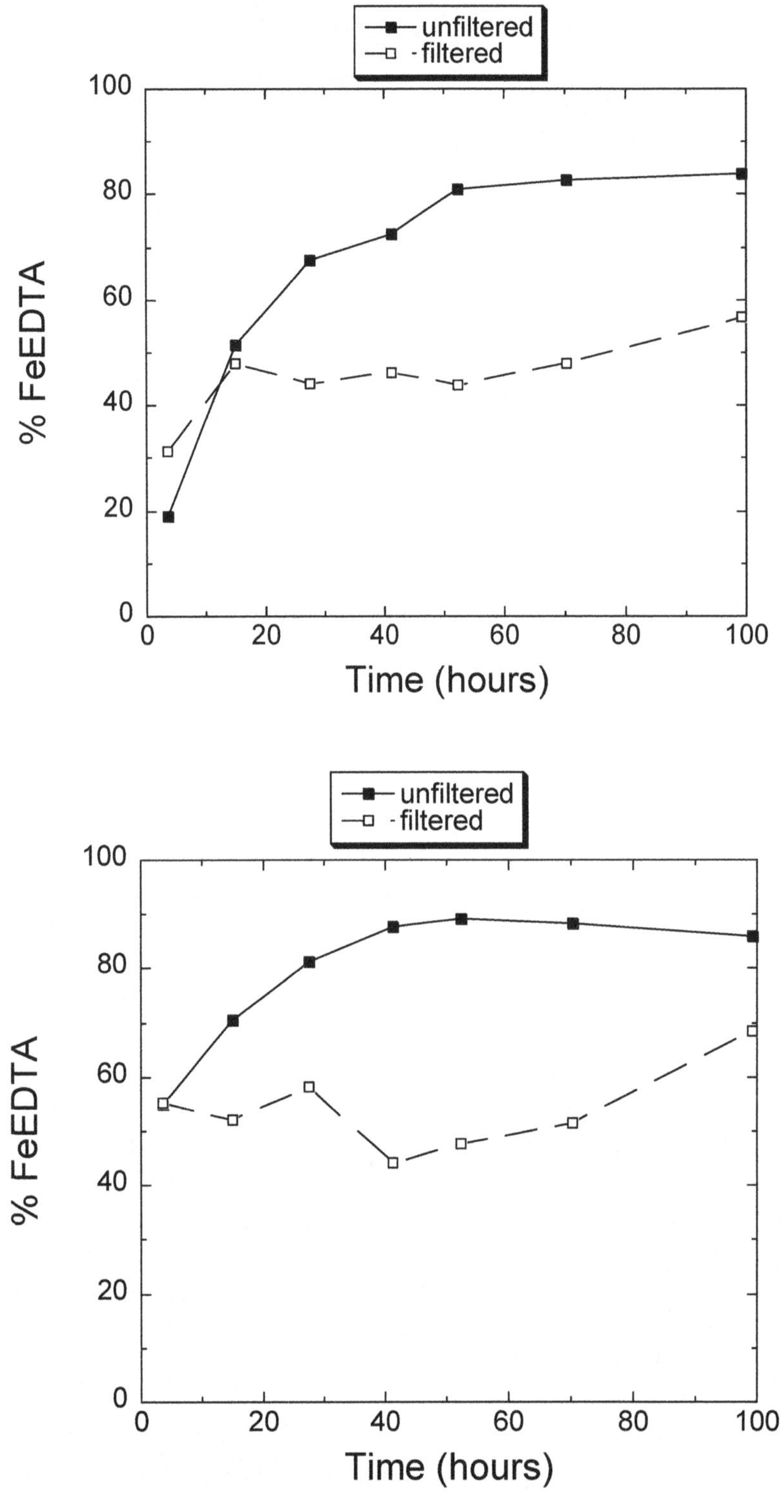


**Figure 4-6. Percentage of EDTA Measured during Storage of Freshly Collected Samples of Primary Influent.**

### 4.3.4 Copper and Zinc

$FeCl_3$ addition had no observable effect on concentrations of Cu (Figure 4-7) and Zn (Figure 4-8) in the secondary effluent of the Millbrae WWTP. Concentrations of dissolved and total Zn and Cu were comparable during the periods with and without $FeCl_3$ addition in samples of primary effluent and secondary effluent. In the secondary effluent samples, the mean total Zn concentration was 227 ± 14 nM during the period with $FeCl_3$ addition and 226 ± 18 nM during the period without. The mean total Cu concentration was 127 ± 8 nM during the period with $FeCl_3$ addition and 131 ± 10 nM during the period without. If the days of the plant upset (Days 52–66) are excluded from the analyses, the total mean metal concentrations were slightly lower during the period without $FeCl_3$ addition (mean $[Cu]_{total}$= 113±13 nM, mean $[Zn]_{tota}$ = 205±26 nM). However, these differences were not statistically significant.

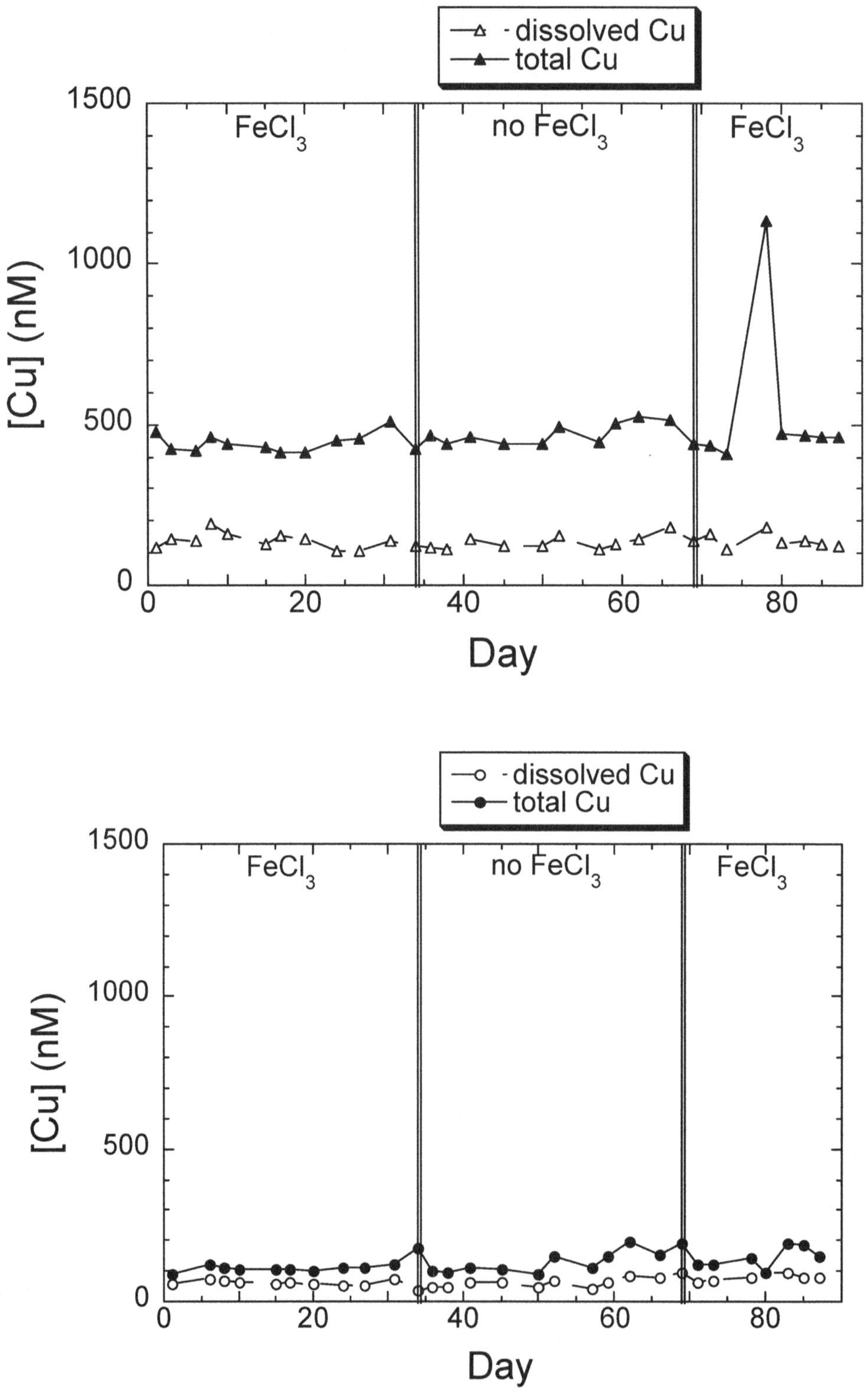


**Figure 4-7. Dissolved and Total Cu Concentrations after Primary (Top) and Secondary (Bottom) Treatment at Millbrae WWTP.**

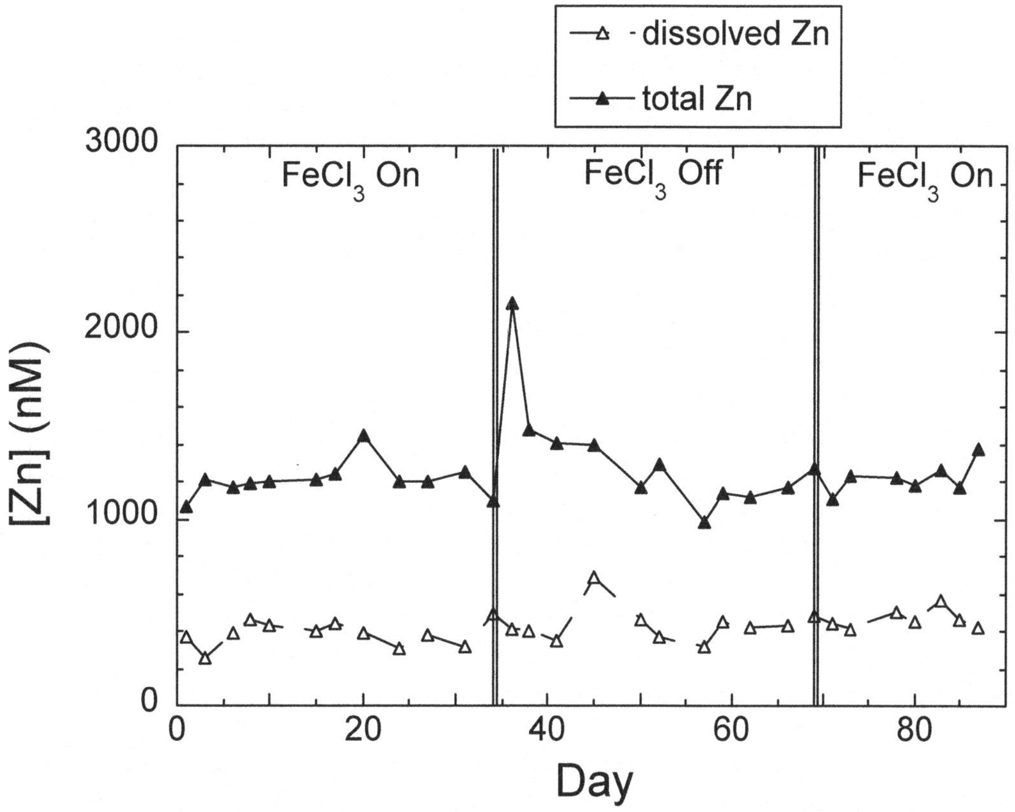


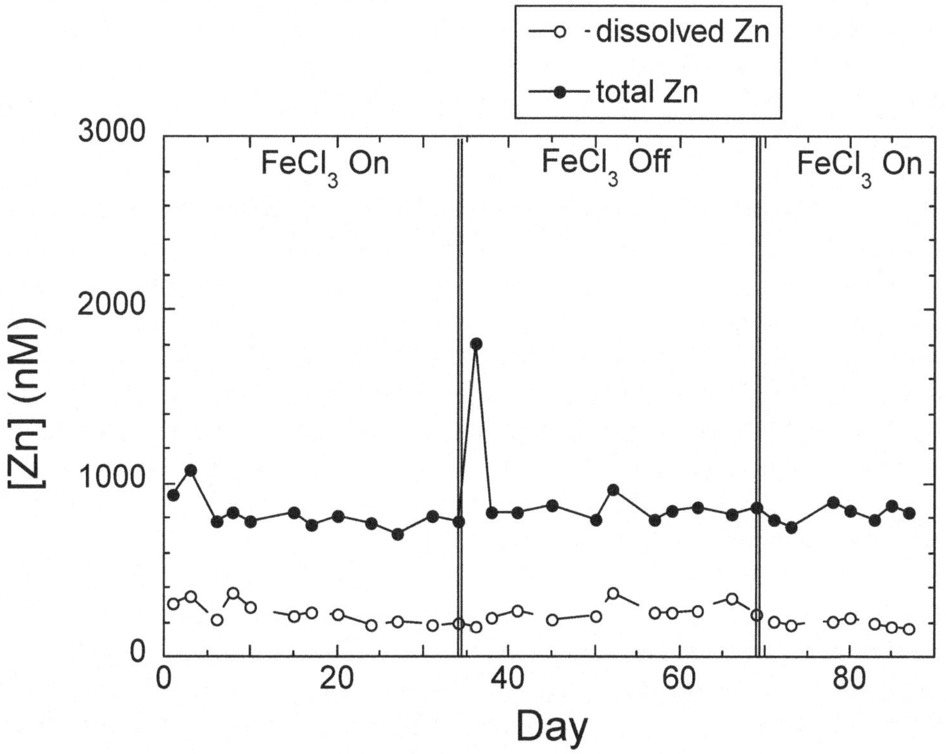


**Figure 4-8. Dissolved and Total Zn Concentrations after Primary (Top) and Secondary (Bottom) Treatment at Millbrae WWTP.**

## 4.4 Interpretation of Results

Measurement of EDTA speciation indicated that pollutant metal-EDTA complexes were converted into FeEDTA complexes during treatment at the Millbrae WWTP (Figure 4-5). Although some of the conversion may have occurred in response to $FeCl_3$ addition, the effect of addition was largely overwhelmed by reactions during primary treatment that were unrelated to $FeCl_3$ addition. The unexpected changes in EDTA speciation during primary treatment are important to metal removal because metal-EDTA complexes often account for a significant fraction of the metals in secondary effluent and any reactions that alter the concentrations of metal-EDTA complexes are likely to affect metal removal.

The observed change in EDTA speciation during primary treatment appears to be related to the presence of iron-containing particles and not to any change in pH that occurs during primary treatment. As shown in Figure 4-6, the percentage of EDTA complexes that consisted of FeEDTA increased in primary influent samples stored in the presence of particles, while little or no change in EDTA speciation was observed in samples that had been filtered with a 0.45-µm filter at the time of sampling. Results of these laboratory experiments indicated that the conversion required several hours, which was consistent with the observed change in EDTA speciation at the Millbrae WWTP, where the percentage of FeEDTA increased by an average of 26% during approximately three hours of primary treatment.

The change in EDTA speciation observed in the laboratory and during primary treatment may have been caused by reactions with particles, beginning in the sewer and continuing during primary settling. Results of synoptic sampling of primary influent and primary effluent during a 10-hour period on October 3, 2002 suggested that the observed changes in EDTA speciation during primary treatment were not an artifact related to collection of grab samples. In pairs of samples corresponding to seven different water packets, the percentage of FeEDTA was approximately 39±5% higher in primary effluent samples than it was in the corresponding primary influent samples (data not shown).

In addition to the conversion of pollutant-metal EDTA complexes that always occurred during primary treatment, some additional conversion of pollutant-metal EDTA complexes into FeEDTA complexes may have occurred when $FeCl_3$ was added. A student t-test comparison of the percentage FeEDTA during the periods with and without $FeCl_3$ addition indicated a nearly significant difference between the means ($p = 0.073$). This apparent change was not due to preferential removal of pollutant-metal EDTA species during the period of $FeCl_3$ addition, because the mean 50 nM decrease in concentration of nonFeEDTA species that occurred during the period of $FeCl_3$ addition was accompanied by a 60 nM increase in the concentration of FeEDTA complexes. The magnitude of the observed conversion is consistent with results from the bench-scale tests described in Chapter 3.0.

Ultimately, the issue of interest to wastewater treatment plant operators is the effect of $FeCl_3$ addition on pollutant metal concentrations in the treatment plant's effluent. There is a general belief among environmental engineers that adding $FeCl_3$ during primary treatment, often accompanied by a low dose of polymer to aid flocculation, will improve pollutant metal removal (Ødegaard, 1998; Shao et al., 1996; Jimenez and Landa, 1998).

While results of the bench-scale study suggested a modest improvement in metal removal may occur after addition of $FeCl_3$, the full-scale study did not indicate a decrease in concentrations of either Cu or Zn in secondary effluent when $FeCl_3$ was added during primary treatment. It is possible that small improvements in metal removal could not be detected due to variability in metal concentrations in grab samples and that the relatively modest dose of $FeCl_3$ employed at the Millbrae WWTP (10 mg $FeCl_3$ $L^{-1}$) was too low to affect metal removal. As indicated by a bench-scale study (Figures 3-6, 3-7), increasing the $FeCl_3$ dose from 15 to 45 mg $FeCl_3$/L improved metal removal. Although $FeCl_3$ doses of up to 40 mg $FeCl_3$/L have been used in wastewater treatment (Shao et al., 1996) higher doses were not used at the Millbrae WWTP because the operators had previously encountered unfavorable changes in floc structure with higher doses of $FeCl_3$.

Although $FeCl_3$ addition was expected to reduce TSS and the concentrations of particle-associated metals in primary effluent, no such effects were observed. Excluding data from the period of plant upset, no difference was seen between mean TSS values in primary effluent during the period of $FeCl_3$ addition as compared to the period without (Figure 4-1). Because $FeCl_3$ addition did not have an observable effect on TSS in primary or secondary effluent, $FeCl_3$ addition did not affect concentrations of particle-associated Cu or Zn entering secondary treatment. Therefore, the mass of metals in the RAS did not decrease when $FeCl_3$ was added.

Despite findings of bench-scale experiments that addition of $FeCl_3$ during simulated primary treatment can convert pollutant-metal EDTA complexes into forms more amenable to treatment, no improvement in metal removal was observed during periods of $FeCl_3$ addition at the Millbrae WWTP. At this plant, the effect of $FeCl_3$ addition was obscured by conversion of pollutant-metal EDTA complexes into FeEDTA complexes during the three hours of primary treatment, which occurred irrespective of $FeCl_3$ addition. As a result of the conversion, it was difficult to effect a change in EDTA speciation as a means of improving metal removal. At other WWTPs, such as those discussed in Chapter 2.0, EDTA is more important to metal speciation. This difference could be related to the relatively long retention time of the primary treatment system at the Millbrae WWTP. Future research may explain the reasons why EDTA conversion was observed at this WWTP. Such information may be useful in future attempts to optimize metal removal in municipal wastewater treatment plants.

# REFERENCES


Alder A.C., Siegrist, H., Gujer, W., Giger, W. *Water Research* **1990**, 24, 733–742.

APHA. Standard Methods for the Examination of Water and Wastewater. American Public Health Association, Washington D.C., 1998.

Bedsworth, W.W. and Sedlak, D.L. *Environmental Science and Technology* **1999**, 33, 926–931.

Bedsworth, W.W. and Sedlak, D.L. *Journal of Chromatography A* **2001**, 109, 57–162.

Buchberger, W., Haddad, P.R., Alexander, P.W. *Journal of Chromatography A*, **1991**, 558, 181-186.

Cheng, M.H., Patterson, J.W., Minear, R.A. *Journal of the Water Pollution Control Federation* **1975**, 47, 362–376.

Cleveland, J.M and Rees, T.F. *Science* **1981**, 212, 1506-1509.

Ekster, A. and Jenkins, D. *Water Environment Research* **1996**, 68, 1172–1178.

Jimenez, C.B. and Landa, V.H. *Water Science and Technology* **1998**, 37, 1–8.

Kari, F.G. and Giger, W. *Water Research* **1996**, 30(1), 122–134.

Means, J.L. and Crerar, D.A. *Science* **1978**, 200, 1477-1481.

Nowack, B., Kari, F.G., Hilger, S.U., Sigg, L. *Anal. Chem.*, **1996**, 68(3), 561-566.

Ødegaard, H., Coagulation as the First Step in Wastewater Treatment. In *Pretreatment in Chemical Water and Wastewater Treatment*. Hahn, H.H. and Klute R., Eds., Springer-Verlag, 1988; pp. 250–260.

Rudd, T., Sterritt, R.M., Lester, J.N. *Water Research* **1984**, 18, 379–384.

Sedlak, D.L., Phinney, J.T., Bedsworth, W.W. *Environmental Science and Technology* **1997**, 31, 3010–3016.

Shao, Y.J., Liu, A., Wada F., Crosse J., Jenkins, D. *Water Science and Technology* **1996**, 34, 223–233.

Water Environment Research Foundation (WERF). *Assessing Methods of Removing Metals from Wastewater: A Review of Data and Methodologies*. Water Environment Research Foundation, Alexandria, VA, 2000.

Ye, L.W. and Lucy C.A. *Anal. Chem.* **1995**, 67, 2534-2538.

WASTEWATER UTILITY

### Alabama
Montgomery Water Works & Sanitary Sewer Board

### Alaska
Anchorage Water & Wastewater Utility

### Arizona
Gila Resources

Glendale, City of, Utilities Department

Mesa, City of

Peoria, City of

Pima County Wastewater Management

### Arkansas
Little Rock Wastewater Utility

### California
Calaveras County Water District

Central Contra Costa Sanitary District

Contra Costa Water District

Crestline Sanitation District

Delta Diablo Sanitation District

Dublin San Ramon Services District

East Bay Dischargers Authority

East Bay Municipal Utility District

El Dorado Irrigation District

Fairfield-Suisun Sewer District

Fresno Department of Public Utilities

Irvine Ranch Water District

Las Virgenes Municipal Water District

Livermore, City of

Lodi, City of

Los Angeles, City of

Los Angeles County, Sanitation Districts of

Napa Sanitation District

Orange County Sanitation District

Palo Alto, City of

Riverside, City of

Sacramento Regional County Sanitation District

San Diego Metropolitan Wastewater Department, City of

San Francisco, City & County of

San Jose, City of

Santa Barbara, City of

Santa Rosa, City of

South Bayside System Authority

South Coast Water District

South Orange County Wastewater Authority

Sunnyvale, City of

Union Sanitary District

West Valley Sanitation District

### Colorado
Boulder, City of

Colorado Springs Utilities

Littleton/Englewood Water Pollution Control Plant

Metro Wastewater Reclamation District, Denver

### Connecticut
The Mattabasset District

New Haven, City of, WPCA

### District of Columbia
District of Columbia Water & Sewer Authority

### Florida
Broward, County of

Fort Lauderdale, City of

Gainesville Regional Utilities

JEA

Miami-Dade Water & Sewer Authority

Orange County Utilities Department

Orlando, City of

Reedy Creek Improvement District

Seminole County Environmental Services

St. Petersburg, City of

Stuart Public Utilities

Tallahassee, City of

Tampa, City of

Toho Water Authority

West Palm Beach, City of

### Georgia
Atlanta Department of Watershed Management

Augusta, City of

Clayton County Water Authority

Cobb County Water System

Columbus Water Works

Fulton County

Gwinnett County Department of Public Utilities

Savannah, City of

### Hawaii
Honolulu, City and County of

### Illinois
American Bottoms Wastewater Treatment Plant

Greater Peoria Sanitary District

Kankakee River Metropolitan Agency

Metropolitan Water Reclamation District of Greater Chicago

North Shore Sanitary District

Wheaton Sanitary District

### Iowa
Ames, City of

Cedar Rapids Wastewater Facility

Des Moines Metro Wastewater Reclamation Authority

Iowa City, City of

### Kansas
Johnson County Unified Wastewater Districts

Unified Government of Wyandotte County/Kansas City, City of

### Kentucky
Louisville & Jefferson County Metropolitan Sewer District

### Louisiana
Sewerage & Water Board of New Orleans

### Maine
Bangor, City of

Portland Water District

### Maryland
Anne Arundel County Bureau of Utility Operations

Howard County Department of Public Works

Washington Suburban Sanitary Commission

### Massachusetts
Boston Water & Sewer Commission

Upper Blackstone Water Pollution Abatement District

### Michigan
Ann Arbor, City of

Detroit, City of

Holland Board of Public Works

Lansing, City of

Owosso Mid-Shiawassee County WWTP

Saginaw, City of

Wayne County Department of Environment

Wyoming, City of

### Minnesota
Rochester, City of

Western Lake Superior Sanitary District

### Missouri
Independence, City of

Kansas City Missouri Water Services Department

Little Blue Valley Sewer District

### Nebraska
Lincoln Wastewater System

### Nevada
Henderson, City of

### New Jersey
Bergen County Utilities Authority

Passaic Valley Sewerage Commissioners

### New York
New York City Department of Environmental Protection

Rockland County Solid Waste Management Authority/ Sewer District

### North Carolina
Charlotte/Mecklenburg Utilities

Durham, City of

Metropolitan Sewerage District of Buncombe County

Orange Water & Sewer Authority

### Ohio
Akron, City of

Butler County Department of Environmental Services

Columbus, City of

Metropolitan Sewer District of Greater Cincinnati

Northeast Ohio Regional Sewer District

Summit, County of

### Oklahoma
Tulsa, City of

### Oregon
Clean Water Services

Eugene/Springfield Water Pollution Control

Water Environment Services

### Pennsylvania
Philadelphia, City of

University Area Joint Authority, State College

### South Carolina
Charleston Commissioners of Public Works

Mount Pleasant Waterworks & Sewer Commission

Spartanburg Sanitary Sewer District

### Tennessee
Cleveland, City of

Knoxville Utilities Board

Murfreesboro Water & Sewer Department

Nashville Metro Water Services

### Texas
Austin, City of

Dallas Water Utilities

Denton, City of

El Paso Water Utilities

Fort Worth, City of

Gulf Coast Waste Disposal Authority

Houston, City of

San Antonio Water System

Trinity River Authority

### Utah
Salt Lake City Corporation

### Virginia
Alexandria Sanitation Authority

Arlington, County of

Fairfax County Virginia

Hampton Roads Sanitation District

Henrico, County of

Hopewell Regional Wastewater Treatment Facility

Loudoun County Sanitation Authority

Lynchburg Regional WWTP

Prince William County Service Authority

Richmond, City of

Rivanna Water & Sewer Authority

### Washington
Edmonds, City of

Everett, City of

King County Department of Natural Resources

Seattle Public Utilities

Sunnyside, Port of

Yakima, City of

### Wisconsin
Green Bay Metro Sewerage District

Madison Metropolitan Sewerage District

Milwaukee Metropolitan Sewerage District

Racine, City of

Sheboygan Regional Wastewater Treatment

Wausau Water Works

### Australia
South Australian Water Corp.

Sydney Water Corp.

Water Corp. of Western Australia

### Canada
Greater Vancouver Regional District

Toronto, City of, Ontario

Winnipeg, City of, Manitoba

### Mexico
Servicios de Agua y Drenaje de Monterrey, I.P.D.

### New Zealand
Watercare Services Limited

### Singapore
Singapore Public Utilities Board

### United Kingdom
Yorkshire Water Services Limited

## STORMWATER UTILITY

### California
Los Angeles, City of, Department of Public Works

Monterey, City of

San Francisco, City & County of

Santa Rosa, City of

Sunnyvale, City of

### Colorado
Boulder, City of

### Georgia
Griffin, City of

### Iowa
Cedar Rapids Wastewater Facility

Des Moines Metro Wastewater Reclamation Authority

### Kansas
Overland Park, City of

### Kentucky
Louisville & Jefferson County Metropolitan Sewer District

### Maine
Portland Water District

### Minnesota
Western Lake Superior Sanitary District

### North Carolina
Charlotte, City of, Stormwater Services

### Pennsylvania
Philadelphia, City of

### Tennessee
Chattanooga Storm Water Management

### Washington
Seattle Public Utilities

## STATE

Arkansas Department of Environmental Quality

Fresno Metropolitan Flood Control District, Calif.

Urban Drainage & Flood Control District, Colo.

## CORPORATE

ADS Environmental Services

The ADVENT Group Inc.

Alan Plummer & Associates

Alden Research Laboratory

Aqua-Aerobic Systems, Inc.

Aquateam–Norwegian Water Technology Centre A/S

BaySaver Inc.

BioVir Laboratories, Inc.

Black & Veatch

Boyle Engineering Corporation

Brown & Caldwell

Burns & McDonnell

CABE Associates Inc.

The Cadmus Group

Camp Dresser & McKee Inc.

Carollo Engineers Inc.

Carpenter Environmental Associates Inc.

CDS Technologies Inc.

Chemtrac Systems Inc.

CH2M HILL

Damon S. Williams Associates, LLC

David L. Sheridan, P.C.

Dewling Associates, Inc.

Earth Tech Inc.

Eco-Matrix

Ecovation

Environmental Engineers International

EMA Inc.

The Eshelman Company Inc.

Finkbeiner, Pettis, & Strout Inc.

Freese & Nichols, Inc.

ftn Associates Inc.

Fuss & O'Neill, Inc.

Gannett Fleming Inc.

Golder Associates Inc.

Greeley and Hansen LLC

Hazen & Sawyer, P.C.

HDR Engineering Inc.

HNTB Corporation

HydroQual Inc.

Infilco Degremont Inc.

Ingersoll-Rand Energy Systems

Insituform Technologies Inc.

Institute for Environmental Technology & Industry, Korea

Jacobson Helgoth Consultants Inc.

Jason Consultants Inc.

Jordan, Jones, & Goulding Inc.

KCI Technologies Inc.

Kelly & Weaver, P.C.

Kennedy/Jenks Consultants

Komline Sanderson Engineering Corporation

Lawler, Matusky & Skelly Engineers, LLP

Limno-Tech Inc.

Lombardo Associates Inc.

Malcolm Pirnie Inc.

McKim & Creed

MEC Analytical Systems Inc.

Metcalf & Eddy Inc.

MWH

New England Organics

Odor & Corrosion Technology Consultants Inc. (OCTC)

Oswald Green, LLC

PA Government Services Inc.

Parametrix Inc.

Parsons

Post, Buckley, Schuh & Jernigan

R&D Engineering/Conestoga Rover & Associates

The RETEC Group

RMC, Inc. (Raines, Melton & Carella)

R.M. Towill Corporation

Ross & Associates Ltd.

Royce Technologies

SAIC Maritime Technical Group

Short Elliott Hendrickson, Inc.

Stantec Consulting Group, Inc.

Stormwater Management, Inc.

Synagro Technologies, Inc.

Tetra Tech Inc.

Trojan Technologies Inc.

URS Corporation

Veolia Water NATC

Wade-Trim Inc.

Weston Solutions Inc.

Woodard & Curran

WRc/D&B, LLC

WWETCO, LLC

Zoeller Pump Company

## INDUSTRY

American Electric Power

ChevronTexaco Energy Research & Technology Company

The Coca-Cola Company

Dow Chemical Company

DuPont Company

Eastman Kodak Company

Eli Lilly & Company

Merck & Company Inc.

ONDEO Services

Procter & Gamble Company

PSEG Services Corp.

RWE Thames Water Plc

Severn Trent Services Inc.

United Water Services LLC

**Note: List as of 7/1/04**

# WERF Product Order Form

As a benefit of joining the Water Environment Research Foundation, subscribers are entitled to receive one complimentary copy of all final reports and other products.Additional copies are available at cost (usually $10).To order your complimentary copy of a report, please write "free" in the unit price column.WERF keeps track of all orders.If the charge differs from what is shown here, we will call to confirm the total before processing.

Name ____________________ Title ____________________

Organization ____________________

Address ____________________

City ____________ State ____________ Zip Code ____________ Country ____________

Phone ____________ Fax ____________ Email ____________

| Stock # | Product | Quantity | Unit Price | Total |
|---|---|---|---|---|
| | | | | |
| | | | | |
| | | | | |
| | | | | |
| | | | | |
| | | | | |
| | | | | |
| | | | | |
| | | | Postage & Handling | |
| | | | VA Residents Add 4.5% Sales Tax | |
| | | | Canadian Residents Add 7% GST | |
| | | | TOTAL | |

## Method of Payment:

(All orders must be prepaid.)

- ☐ Check or Money Order Enclosed
- ☐ Visa ☐ Mastercard ☐ American Express

Account No. ____________ Exp. Date ____________

Signature ____________

## Shipping & Handling:

| Amount of Order | United States | Canada & Mexico | All Others |
|---|---|---|---|
| Up to but not more than: | Add: | Add: | Add: |
| $20.00 | $5.00* | $8.00 | 50% of amount |
| 30.00 | 5.50 | 8.00 | 40% of amount |
| 40.00 | 6.00 | 8.00 | |
| 50.00 | 6.50 | 14.00 | |
| 60.00 | 7.00 | 14.00 | |
| 80.00 | 8.00 | 14.00 | |
| 100.00 | 10.00 | 21.00 | |
| 150.00 | 12.50 | 28.00 | |
| 200.00 | 15.00 | 35.00 | |
| More than $200.00 | Add 20% of order | Add 20% of order | |

*minimum amount for all orders

Note: Please make checks payable to the Water Environment Research Foundation.

## To Order (Subscribers Only):

Log on to www.werf.org and click on the "Product Catalog."

Phone: (703) 684-2470
Fax: (703) 299-0742.

WERF
Attn: Subscriber Services
635 Slaters Lane
Alexandria, VA 22314-1177

## To Order (Non-Subscribers):

Non-subscribers may be able to order WERF publications either through WEF (www.wef.org) or IWAP (www.iwapublishing.com).Visit WERF's website at www.werf.org for details.

IWA Publishing's authorised EU representative for General Product Safety Regulations is Diane D'Arras, 15 rue Duret, 75116 Paris, France, e-mail: safety@iwap.co.uk.

Printed and bound by CPI Group (UK) Ltd, Croydon, CR0 4YY
17/09/2026
02218793-0013